Lecture Notes in Physics
Monographs

Springer
Berlin
Heidelberg
New York
Barcelona
Hong Kong
London
Milan
Paris
Singapore
Tokyo

Physics and Astronomy

ONLINE LIBRARY

http://www.springer.de/phys/

The Editorial Policy for Monographs

The series Lecture Notes in Physics reports new developments in physical research and teaching - quickly, informally, and at a high level. The type of material considered for publication in the monograph Series includes monographs presenting original research or new angles in a classical field. The timeliness of a manuscript is more important than its form, which may be preliminary or tentative. Manuscripts should be reasonably self-contained. They will often present not only results of the author(s) but also related work by other people and will provide sufficient motivation, examples, and applications.

The manuscripts or a detailed description thereof should be submitted either to one of the series editors or to the managing editor. The proposal is then carefully refereed. A final decision concerning publication can often only be made on the basis of the complete manuscript, but otherwise the editors will try to make a preliminary decision as definite as they can on the basis of the available information.

Manuscripts should be no less than 100 and preferably no more than 400 pages in length. Final manuscripts should be in English. They should include a table of contents and an informative introduction accessible also to readers not particularly familiar with the topic treated. Authors are free to use the material in other publications. However, if extensive use is made elsewhere, the publisher should be informed. Authors receive jointly 30 complimentary copies of their book. They are entitled to purchase further copies of their book at a reduced rate. No reprints of individual contributions can be supplied. No royalty is paid on Lecture Notes in Physics volumes. Commitment to publish is made by letter of interest rather than by signing a formal contract. Springer-Verlag secures the copyright for each volume.

The Production Process

The books are hardbound, and quality paper appropriate to the needs of the author(s) is used. Publication time is about ten weeks. More than twenty years of experience guarantee authors the best possible service. To reach the goal of rapid publication at a low price the technique of photographic reproduction from a camera-ready manuscript was chosen. This process shifts the main responsibility for the technical quality considerably from the publisher to the author. We therefore urge all authors to observe very carefully our guidelines for the preparation of camera-ready manuscripts, which we will supply on request. This applies especially to the quality of figures and halftones submitted for publication. Figures should be submitted as originals or glossy prints, as very often Xerox copies are not suitable for reproduction. For the same reason, any writing within figures should not be smaller than 2.5 mm. It might be useful to look at some of the volumes already published or, especially if some atypical text is planned, to write to the Physics Editorial Department of Springer-Verlag direct. This avoids mistakes and time-consuming correspondence during the production period.

As a special service, we offer free of charge LaTeX and TeX macro packages to format the text according to Springer-Verlag's quality requirements. We strongly recommend authors to make use of this offer, as the result will be a book of considerably improved technical quality.

For further information please contact Springer-Verlag, Physics Editorial Department II, Tiergartenstrasse 17, D-69121 Heidelberg, Germany.

Series homepage – http://www.springer.de/phys/books/lnpm

Nir Polonsky

Supersymmetry: Structure and Phenomena

Extensions of the Standard Model

 Springer

Author

Nir Polonsky
Massachusetts Institute of Technology (MIT)
Center for Theoretical Physics
77 Massachusetts Avenue
Cambridge, MA 02139, USA

Library of Congress Cataloging-in-Publication Data applied for.

Die Deutsche Bibliothek - CIP-Einheitsaufnahme
Polonsky, Nir:
Supersymmetry: structure and phenomena : extensions of the standard model /
Nir Polonsky. - Berlin ; Heidelberg ; New York ; Barcelona ; Hong Kong ;
London ; Milan ; Paris ; Singapore ; Tokyo : Springer, 2001
  (Lecture notes in physics : N.s. M, Monographs ; 68)
  (Physics and astronomy online library)
  ISBN 3-540-42442-3

ISSN 0940-7677 (Lecture Notes in Physics. Monographs)
ISBN 3-540-42442-3 Springer-Verlag Berlin Heidelberg New York

Springer-Verlag Berlin Heidelberg New York
a member of BertelsmannSpringer Science+Business Media GmbH

http://www.springer.de

© Springer-Verlag Berlin Heidelberg 2001
Printed in Germany

Typesetting: Camera-ready by the author
Cover design: *design & production*, Heidelberg

Printed on acid-free paper
SPIN: 10847496   55/3141/du - 5 4 3 2 1 0

To Miriam and Pinhas, to Janice, and most of all, to Nevo

להורי מרים ופנחס , לאליענה , ובמיוחד לנבו

Foreword

Supersymmetry is one of the most discussed themes in modern particle physics, although there is no convincing evidence yet for its existence in nature. Intrigued by the uniqueness and theoretical beauty of this symmetry, researchers in the field are confident that interest will continue in the coming years and even decades. In fact, many of the most exciting recent developments in mathematical physics, quantum field theory and string theory were obtained using the concepts of supersymmetry.

Still the most fascinating aspect of supersymmetry resides in the possibility that it might extend the celebrated standard model of elementary particle physics into an energy range accessible to present and planned experiments: its experimental discovery might thus be "around the corner". This so-called "low-energy" supersymmetry, the supersymmetric generalization of the standard model of strong, weak and electromagnetic interactions, is the central subject of the present book.

Nir Polonsky is well suited as an author of such a book as he is an expert in the field and, throughout his career, has made important contributions to its development. In the present monograph he provides a "guided tour" through this vast field of "low-energy" supersymmetry. In a convincing way he selects the essential topics and provides a consistent account of the field, avoiding the discussion of unnecessary details. His clear and patient explanations make this book useful both for the experienced researcher in the field and the student who wants to take the first steps towards learning supersymmetry. All physicists interested in this subject will want to have this book on their shelves.

Bonn, Germany,
July 2001

Hans Peter Nilles

Preface

Moor: point
my horse
where birds sing.

Basho (Translation by Lucien Stryk)

These notes provide an introductory yet comprehensive discussion of the phenomenology associated with supersymmetric extensions of the Standard Model of electroweak and strong interactions. The choice of topics and of their presentation is meant to draw readers with various backgrounds and research interests, and to provide a pedagogical tour of supersymmetry, structure and phenomena. The intensive and ongoing efforts to understand and discover such phenomena, or to alternatively falsify the paradigm predicting it, and the central role of supersymmetry in particle physics studies, mean that such a journey is a well-motivated endeavor, if not a necessity. This is particularly true as we look forward to moving into a new era of discoveries with the forthcoming commissioning of CERN's Large Hadron Collider (LHC), which is scheduled to operate by the year 2005.

Even though our focus is on (weak-scale) supersymmetry, it would still be impossible to cover within these notes all that it entails. In the following, we choose to stress the more elementary details over the technicalities, the physical picture over its formal derivation. While some issues are covered in depth, many others are only sketched or referred to, leaving the details to either optional exercises or further reading. Our mission is to gain acquaintance with the fundamental picture (regardless of the reader's background), and we will attempt to do so in an intuitive fashion while providing a broad and honest review of the related issues. Many reviews, lecture notes, and text books (most notably, Weinberg's book [1]) are available and references will be given to some of those, as well as to some early and recent research papers on the subjects discussed. Some areas are particularly well covered by reviews, e.g., the algebra on the one hand and experimental searches on the other hand, and hence will not be explored extensively here.

Before discussing supersymmetry and the "TeV World", it is useful to recall and establish our starting point: the Standard Model of electroweak and strong interactions as characterized and tested at currently available energies $\sqrt{s} \lesssim 200 - 300$ GeV (per parton). We therefore begin with a brief summary of the main ingredients of the SM. Hints for an additional structure within experimental reach will be emphasized, and the known theoretical possibilities for such a structure, particularly supersymmetry, will be introduced. This will be done in the first part of the notes. The second part provides a phenomenological "bottom-up" construction of supersymmetry, followed by a brief discussion of the formalism – providing intuitive understanding as well as the necessary terminology. We then supersymmetrize the Standard Model, paying attention to some of the details while only touching upon others.

In the third part of the notes we turn to an exploration of a small number of issues which relate ultraviolet and infrared structures. These include renormalization and radiative symmetry breaking, unification, and the lightness of the Higgs boson, issues which underlie the realization of the weak scale. This particular choice of topics does not only lay the foundation for supersymmetry phenomenology but also represents some of the successful predictions of weak-scale supersymmetry. Subsequently, we discuss in the fourth part some of the difficulties in extending the standard model, the most important of which is the flavor problem. Its possible resolutions are used as an organizing principle for the model space, which is then characterized and described.

Throughout these notes we attempt to leave the reader with a clear view of current research directions and avenues which stem from the above topics. Some more technical and specialized issues are explored further in the fifth and last part of the notes. These include the possible origins of neutrino mass and mixing and the implications of the different scenarios: vacuum stability and the respective constraints; classifications of supersymmetry-breaking operators; and an introductory discussion of extended supersymmetry. These serve as a (subjective) sample of contemporary and potential avenues of research. The reader is encouraged to explore the literature given throughout the notes for technical discussions of other topics of interest.

As physicists, our interest is in confronting theory with experiment. Here, we will argue for and motivate supersymmetry as a reasonable and likely perturbative extension of the standard model. Whether supersymmetry is realized in nature, and if so in what form, can and must be determined by experiment. Experimental efforts in this direction are ongoing and will hopefully result in an answer or a clear indication within the next decade. Discovery will only signal the start of a new era dedicated to deciphering the ultraviolet theory from the data: supersymmetry, by its nature (and as we will demonstrate), relates the infrared and the ultraviolet regimes. By doing so it provides us with extraordinary opportunities for a glimpse at scales we cannot reach, and to answer fundamental questions regarding unification, gravity, and much more. The more one is familiar with the ultraviolet possibilities and their infrared implications, the more one can direct experiments towards a possible discovery.

Clearly, a lot of work is yet to be done and there is always room for new ideas to be put forward. It is my hope that these notes convince and enable one to follow (if not participate in) the story of weak-scale supersymmetry, and new physics in general, as it unveils. It is not our aim to equip the reader with calculation tools or to review in these notes each and every aspect or possibility. Rather, we will attempt to enable the interested reader to form an educated opinion and identify areas of interest for further study and exploration. Many of the issues discussed will be presented from an angle which differs from other presentations in the literature, for the potential benefit of both the novice and expert readers. Our bibliography is comprehensive, however it is far from exhaustive or complete. Most research articles on the subject (which were written in the last decade) can be found in the Los Alamos National Laboratory electronic archive http://xxx.lanl.gov/archive/hep-ph, and the Stanford Linear Accelerator Center's Spires electronic database http://www.slac.stanford.edu/spires/hep lists relevant articles and reviews.

The core of the manuscript is based on the lecture series "Essential Supersymmetry" which appeared in the proceedings of the TASI-98 school [2]. However, the material was revised, updated and significantly extended. I thank Francesca Borzumati for her help with the figures, and Howard Haber and Chris Kolda for providing me with Fig. 8.1 and Fig. 7.2, respectively. I also thank Francesca Borzumati, Lee Hyun Min, Yasanury Nomura, and Myck Schwetz for their careful reading of earlier versions of the manuscript and for their comments. I benefited from work done in collaboration with Jonathan Bagger, Francesca Borzumati, Hsin-Chia Cheng, Jens Erler, Glennys Farrar, Jonathan Feng, Hong-Jian He, Jean-Loic Kneur, Chris Kolda, Paul Langacker, Hans-Peter Nilles, Stefan Pokorski, Alex Pomarol, Shufang Su, Scott Thomas, Jing Wang, and Ren-Jie Zhang. This manuscript was completed during my tenure as a Research Scientist at the Massachusetts Institute of Technology, and I thank Bob Jaffe for his encouragement. Finally, I thank Peter Nilles for his support.

Cambridge, May 2001   *Nir Polonsky*

Contents

**Part III. Supersymmetry Top-Down:
Understanding the Weak Scale**

Part I

Stepping Beyond the Standard Model

1. Basic Ingredients

The Standard Model of strong and electroweak interactions (SM) was extensively outlined and discussed by many, for example in Altarelli's TASI lectures [3] where the reader can find an in-depth discussion and references. The SM has provided the cornerstone of elementary particle physics for nearly three decades, particularly so after it was soundly established in recent years by the various high-energy collider experiments. Even though the reader's familiarity with the SM is assumed throughout these notes, let us recall some of its main ingredients, particularly those that force one to look for its extensions.

Indeed, the most direct evidence for an extended structure was provided most recently by experimentally establishing neutrino flavor oscillations [4]. (See Langacker for details and references [5] as well as recent reviews by Barger [6].) However, let us refrain from discussing neutrino mass and mixing until the last part of these notes where neutrino mass generation within supersymmetric frameworks will be explored. With the exception of Chap. 11, we will adopt the SM postulation of massless neutrinos. More intrinsic and fundamental indications for additional structure (that will ultimately have to also explain the neutrino spectrum) stem from the SM (so-far untested) postulation of a scalar field, the Higgs field, whose vacuum expectation value (*vev*) spontaneously breaks the SM gauge symmetry.

We begin in this chapter with a brief review of the boson and fermion structures in the SM. This summary sets the foundation for the discussion of indications for additional structure, which follows in the next chapter.

1.1 Bosons and Fermions

The SM is a theory of fermions and of gauge bosons mediating the $SU(3)_c \times SU(2)_L \times U(1)_Y$ color and electroweak gauge interactions of the fermions. However, the $SU(2)_L \times U(1)_Y$ weak isospin × hypercharge electroweak symmetry is spontaneously broken and is not respected by the vacuum, as is manifested in its massive W and Z gauge bosons. The spontaneous symmetry breaking is parameterized by a complex scalar field, the Higgs field, which provides the Goldstone bosons of electroweak symmetry breaking. Its inclusion in the weak-scale spectrum restores unitarity and perturbative consistency of WW scattering, for example. Its gauge-invariant Yukawa interactions allow one to simultaneously explain, at least technically so, the $SU(2)_L \times U(1)_Y$ breaking and the chiral-symmetry violating fermion mass spectrum. The SM employs the by-now standard quantum-field theory tools, and it commutes with, rather than incorporates, gravity.

Table 1.1 lists the matter, Higgs and gauge fields which constitute the SM particle (field) content. (The graviton is listed for completeness.) The table also serves to establish the relevant notation. The fermion flavor index $a = 1, 2, 3$ indicates the three identical generations (or families) of fermions which are distinguished only by their mass spectrum. It is important to note that the $U(1)$ hypercharge is *de facto* assigned such that each family consti- tutes an anomaly-free set of fermions. All fields (or their linear combinations corresponding to the physical eigenstates : $B, W \to \gamma, Z, W^\pm$; see below) aside from the Higgs boson have been essentially observed. (Indirect evi- dence for the existence of the Higgs boson from precision measurements of electroweak observables, however, has been recently acquiring statistical sig- nificance.) It is therefore not surprising that it is the symmetry breaking sector which is the least understood. It will occupy most of our attention in the remaining of Part I.

Given the charge assignments, it is straightforward to write down the gauge invariant scalar potential and Yukawa interactions. We begin with the scalar potential and reserve the discussion of Yukawa interactions and flavor to the next section.

The scalar potential is given by

$$V(H) = -m^2 H H^\dagger + \lambda (H H^\dagger)^2, \qquad (1.1)$$

where $\lambda > 0$ is an arbitrary quartic coupling. If $m^2 < 0$ then the global minimum of the potential is at the origin and all symmetries are preserved. However, for $m^2 > 0$, as we shall assume, the Higgs doublet acquires a non- vanishing *vev* $\langle H \rangle = (\nu, 0)/\sqrt{2}$ where $\nu \equiv \sqrt{(m^2/\lambda)}$.

The electroweak $SU(2)_L \times U(1)_Y$ symmetry is now spontaneously broken down to $U(1)_Q$ of QED with $Q = T_3 + Y/2$. (A $SU(2)$ rotation was used to fix the Higgs *vev* in its conventionally chosen direction). The charged and neutral CP-odd Higgs components are Goldstone bosons in this case and are absorbed by the electroweak gauge bosons which, as given by experiment, are massive with [7] $M^2_{W^\pm} = g^2\nu^2/4 = (80.419 \pm 0.056 \text{ GeV})^2$ and $M^2_Z = (g^2 + g'^2)\nu^2/4 = (91.1882 \pm 0.0022 \text{ GeV})^2$. Here, $g \equiv g_2$ and g' are the conventional notation for the $SU(2)$ and hypercharge gauge couplings, respectively. (Note that unlike the symbols g and g_2 which will be used interchangeably, the symbols g' used here and g_1 used in Chap. 6 differ in their normalization. The symbol g will be also used to denote a gauge coupling in general.) One defines the weak angle $\tan\theta_W \equiv g'/g$ so that $Z = \cos\theta_W W_3 - \sin\theta_W B$, and the massless photon of the unbroken QED is given by the orthogonal combination. (The charged mass eigenstates are $W^\pm = (W_1 \mp iW_2)/\sqrt{2}$.) The QED coupling is $e = g\sin\theta_W$. From the Fermi constant, $G_F = g^2/4\sqrt{2}M^2_W$, measured in muon decay, one can extract $\nu = (\sqrt{2}G_F)^{-1/2} = 246$ GeV. (Our normalization is the conventional one for the one-Higgs doublet SM and will be modified when discussing the two-Higgs doublet supersymmetric extensions.)

Note the (tree-level) mass relation $M_W^2 = M_Z^2 \cos^2 \theta_W$. We will mention below the quantum corrections to this relation, measured by the ρ-parameter, $M_W^2 \equiv \rho M_Z^2 \cos^2 \theta_W$. In fact $\rho = 1$ at tree level in the case of Higgs doublets but not, for example, in the case of Higgs triplets. Its fitted value (subtracting SM quantum effects $\sim \mathcal{O}(m_t^2)$), $\rho \simeq 1$ [7], therefore severely constrains the possibility of an electroweak breaking vev of a $SU(2)_L$-triplet Higgs field. In particular, the SM, as well as any of its extensions, must assume that the Higgs field(s) of electroweak symmetry breaking is a (are) $SU(2)_L$-doublet(s).

Table 1.1. The SM field content. Our notation $(Q_c, Q_L)_{Q_{Y/2}}$ lists color, weak isospin and hypercharge assignments of a given field, respectively, and $Q_c = 1$, $Q_L = 1$ or $Q_{Y/2} = 0$ indicate a singlet under the respective group transformations. The T_3 isospin operator is $+1/2$ $(-1/2)$ when acting on the upper (lower) component of an isospin doublet (and zero otherwise).

Sector	Spin	Field
$SU(3)$ gauge bosons (gluons)	1	$g \equiv (8, 1)_0$
$SU(2)$ gauge bosons		$W \equiv (1, 3)_0$
$U(1)$ gauge boson		$B \equiv (1, 1)_0$
Chiral matter (Three families: $a = 1, 2, 3$)	$\frac{1}{2}$	$Q_a \equiv \begin{pmatrix} u \\ d \end{pmatrix}_{L_a} \equiv (3, 2)_{\frac{1}{6}}$
		$U_a \equiv u^c_{L_a} \equiv (\bar{3}, 1)_{-\frac{2}{3}}$
		$D_a \equiv d^c_{L_a} \equiv (\bar{3}, 1)_{\frac{1}{3}}$
		$L_a \equiv \begin{pmatrix} \nu \\ e^- \end{pmatrix}_{L_a} \equiv (1, 2)_{-\frac{1}{2}}$
		$E_a \equiv (e^-)^c_{L_a} \equiv (1, 1)_1$
Symmetry breaking (the Higgs boson)	0	$H \equiv \begin{pmatrix} H^0 \\ H^- \end{pmatrix} \equiv (1, 2)_{-\frac{1}{2}}$
Gravity (the graviton)	2	$G \equiv (1, 1)_0$

1.2 Flavor

The gauge invariant Yukawa interactions are of the from $\psi_L H \psi_R$ or of the form $\psi_L i\sigma_2 H^* \psi_R$, where we labeled the (left-handed) chiral matter transforming under $SU(2)_L$ and the (right-handed) chiral matter which is a weak-isopsin singlet with the L and R subscripts, respectively. One has

$$\mathcal{L}_{\text{Yukawa}} = y_{l_{ab}} L_a H E_b + y_{d_{ab}} Q_a H D_b + y_{u_{ab}} Q_a (i\sigma_2 H^*) U_b + h.c., \qquad (1.2)$$

where $SU(2)$ indices are implicit and are contracted with the antisymmetric tensor ϵ_{ij}: $\epsilon_{12} = +1 = -\epsilon_{21}$. Note that the choice of the hypercharge sign for H is somewhat arbitrary since one can always define $\bar{H} = i\sigma_2 H^*$ which carries the opposite hypercharge (σ_2 is the Pauli matrix and $\bar{H} = (H^+, -H^{0*})^T$ given our choice). This will not be the case in the supersymmetric extension. One can rewrite the Lagrangian (1.2) in terms of the physical CP-even component η (with mass $\sqrt{2}m$) and the *vev* of $H(x) \rightarrow ((\nu + \eta(x), 0)/\sqrt{2})^T$ in order to find the physical Higgs Yukawa interactions and the fermion (Dirac) mass terms $m_{f_{ab}} = y_{f_{ab}} \nu/\sqrt{2}$.

It is interesting to note that had the spectrum contained a SM singlet fermion $N \equiv (1,1)_0$ (with a lepton number -1) then a (lepton-number preserving) neutrino Yukawa/mass term $y_{\nu_{ab}} L_a (i\sigma_2 H^*) N_b + h.c.$ could also be written. N is the right-handed neutrino. Being a SM singlet it could also have a gauge-invariant Majorana mass term $\sim N_a N_b$ which violates lepton number by two units. However, the SM contains no right-handed neutrinos, and the (left-handed) neutrinos are assumed massless. Furthermore, lepton L ($L = +1$ for L_a and $L = -1$ for E_a, N_a) and baryon B ($B = +1/3$ for Q_a and $B = -1/3$ for U_a, D_a) numbers are automatically conserved (in perturbation theory) by its interactions, e.g. see eqs. (1.2) and (1.1). L and B are accidental but exact symmetries of the SM, and their conservation holds to all orders in perturbation theory. In particular, the lightest baryon, the proton, is predicted to be stable. We also note in passing that the introduction of a right-handed neutrino N_a (per family) allows for a new anomaly-free gauged $U(1)$ symmetry $U(1)_{Q'}$ with $Q' = B - L$. It would also allow one to arrange the SM right-handed fields in isospin doublets of a $SU(2)_R$ gauge symmetry [8] such that $Y = T_{3_R} + [(B - L)/2]$ (and as before, $Q = T_{3_L} + Y/2$ for QED). Hence, extended gauge symmetries and right-handed neutrinos are naturally linked. (This is particularly relevant in the case of grand-unified theories which are discussed in Chap. 6.) We return to the discussion of neutrinos and lepton number in Chap. 11.

Returning to the SM, both Yukawa couplings and fermion masses are written above as 3×3 matrices in the corresponding (up, down, lepton) flavor space. (The dimension of the flavor space is given by the number of families, $N_f = 3$.) In order to obtain the physical mass eigenstates one has to diagonalize the different Yukawa matrices and subsequently to perform independent unitary transformations on all vectors in flavor space. For example, $u_{a_L} \equiv (u, c, t)_L$ and $d_{a_L} \equiv (d, s, b)_L$ (employing standard flavor symbols which we will use interchangeably with the symbols defined in Table 1.1.) are now rotated by different transformations U and V, respectively, which correspond to the diagonalization of the respective Yukawa matrix. Hence, the weak (interaction) and mass (physical) eigenstates are not identical: $d_{L_a}^W = V^{ab} d_{L_b}^m$ etc. where the superscripts W and m denote weak and mass eigenstates, respectively.

Charged currents $J_{CC} \sim \bar{u}^W \Gamma T d^W + h.c.$ (interacting with the charged gauge bosons $W^{\pm} J_{CC}^{\pm}$) are then proportional, when rotated to the physical mass basis, to $U^{\dagger} V \equiv V_{CKM}$. Here, Γ and T denote Dirac and $SU(2)_L$ matrices, respectively, and subscripts were omitted. (Note that only the product V_{CKM} appears in physical observables. One could choose, for convenience, a basis such that $U = I$ and $V \equiv V_{CKM}$.) The 3×3 Cabibbo-Kobayashi-Maskawa V_{CKM} (unitary) matrix contains three independent angles and one phase parameter and connects the different generations: Charged currents contain flavor off-diagonal interaction vertices which lead to flavor changing currents. For example, $V_{CKM}^{us} \bar{c}^m \Gamma T d^m + h.c. \subset J_{CC}$ leads to flavor changing W decays, $W^+ \to c\bar{d}$.

On the other hand, unitarity guarantees that no flavor changing neutral currents (FCNC) arise (at tree level): $J_{NC} \sim \bar{u}^W \Gamma T u^W + \bar{d}^W \Gamma T d^W + h.c.$ (interacting with the Z and with the photon) is proportional to $U^{\dagger} U, V^{\dagger} V \equiv I$ and is basis invariant. (Loop corrections change the situation due to the flavor non-conserving nature of the charged currents and lead, for example, to loop-induced meson mixing and quark decays such as $b \to s\gamma$.) In addition, the absence of right-handed electroweak-singlet neutrinos and of the corresponding Yukawa terms implies that a flavor rotation of the neutrinos is not physical (since obviously no neutrino Yukawa/mass term is rotated/diagonalized) and therefore can always be chosen so that the left-handed neutrinos align with the charged leptons. Hence no leptonic flavor (*i.e.*, e, μ, τ) violating (LFV) currents, charged or neutral, arise in the SM at any order in perturbation theory. Both predictions provide important arenas for testing extensions of the SM, which often predict new sources of FCNC (which appear only at loop level in the SM) and/or LFV (which are absent in the SM).

1.3 Status of the Standard Model

Even though the SM is extremely successful in explaining all observed phenomena to date, it does not provide any guidelines in choosing its various *a priori* 18 free parameters (in addition to its rank and representations): the gauge couplings, the number of chiral families, the Higgs parameters and Yukawa couplings, and ultimately, the weak scale $\sim g v$.

The fermion mass spectrum is indeed successfully related to the spontaneous gauge symmetry breaking: The same Higgs *vev* which breaks the gauge symmetry also spontaneously breaks the chiral symmetries of the SM (which forbid, if preserved, massive fermions). Nevertheless, the size of the breaking of the chiral symmetries is put in by hand in (1.2). Subsequently, all flavor parameters - fermion masses and CKM angles and phase – have to be fixed by hand: The SM does not guide us as to the origins of flavor (nor as to the family duplication). Charge quantization and the quark-lepton distinction also remain mysterious (even though they can be argued to provide a unique anomaly-free set of fermions, given the gauge symmetries).

One may argue that these (and the exclusion of gravity) are sufficient reasons to view the SM as only a "low-energy" limit of an extended "more fundamental" theory, which may provide some of the answers. However, an even stronger motivation to adopt the effective theory point of view emerges from the discussion of quantum corrections to the Higgs potential (1.1), and in fact, it will restrict the possible frameworks in which one can consistently address all other fundamental questions. We turn to this subject in the next chapter.

2. The Hierarchy Problem and Beyond

2.1 The Hierarchy Problem

Using standard field theory tools, one renormalizes the SM Lagrangian in order to account for quantum (*i.e.*, loop) corrections. Indeed, this procedure is successfully carried out in the case of the extrapolation of "the effective" $SU(3)_c \times U(1)_Q$ theory to the weak scale and its embedding in the $SU(3)_c \times SU(2)_L \times U(1)_Y$ SM, as was described by Altarelli [3]. (See also Fig. 6.1 below.) While most corrections are at most logarithmically divergent and hence correspond a well-understood logarithmic renormalization (and finite shifts[1]) of parameters, there is an important exception. The model now contains the Higgs field whose lightness is not protected by either gauge or chiral symmetries. The leading corrections to the Higgs two-point function (and hence, to its mass parameter in (1.1)) depend quadratically on the ultraviolet cut-off scale. Adopting the view that the SM itself is only an effective low-energy theory, the renormalization procedure is again understood as extrapolating the model towards the more fundamental ultraviolet scale. The quadratic dependence on this scale, however, is forcing the theory to reside at the ultraviolet scale itself[2], hence leading to an apparent paradox with crucial implications. It amounts to a failure to consistently accommodate fundamental scalar fields within the framework of the SM, undermining the notion of an ultraviolet independent low-energy theory.

More specifically, one needs to consider, at one-loop order, the three classes of quadratically divergent one-loop contributions shown in Figs. 2.1–2.3. The first divergence results from the Higgs $SU(2) \times U(1)$ gauge interactions. (Since we are interested in the ultraviolet dependence we will only

---

[1] In fact, from the effective low-energy theory point of view, finite corrections which are proportional to the square of the top mass m_t^2 are quadratically divergent corrections which are cut off by m_t. Such corrections appear at the quantum level, for example, in the neutral to charged current ratio, the ρ-parameter $\sim M_W^2/M_Z^2 \cos^2 \theta_W \sim 1 + \mathcal{O}(m_t^2)$, and also in the $Z \to b\bar{b}$ branching ratio (which is relevant for the effective theory at M_Z).

[2] Again, an analogy with the top quark is in place: The electroweak theory indeed resides near the cut-off on its quadratic divergences $\sim m_t$. (Note that quantum corrections to electroweak observables depend only logarithmically on the Higgs mass, which is assumed to be of the same scale.)

consider the full $SU(2) \times U(1)$ theory above the weak scale.) The gauge bosons and the Higgs fields themselves are propagating in the loops. A naive power counting gives for the corrections to the Higgs mass

$$\delta m^2_{\text{gauge}} \sim C_W \frac{g^2}{16\pi^2} \Lambda^2_{\text{UV}}, \qquad (2.1)$$

where C_I are numerical coefficients whose value is immaterial here. We included only the leading contributions in the ultraviolet cut-off Λ_{UV} (neglecting logarithmic dependences), and similarly for the hypercharge contribution which we omit. A second contribution results from the Higgs Yukawa interactions. In this case fermions are propagating in the loops and one has

$$\delta m^2_{\text{Yukawa}} \sim -\sum_f C_f \frac{y_f^2}{16\pi^2} \Lambda^2_{\text{UV}}, \qquad (2.2)$$

where the summation is over all fermion flavors. (Small Yukawa couplings are not negligible once $\Lambda_{\text{UV}} \to \infty$!) Note the negative sign due to the fermion loop. Lastly, the Higgs self interaction leads to another contribution which is proportional to its quartic coupling,

$$\delta m^2_{\text{Higgs}} \sim C_H \frac{\lambda}{16\pi^2} \Lambda^2_{\text{UV}}. \qquad (2.3)$$

Summing all independent contributions one finds

$$-m^2_{\text{tree}} + \frac{1}{16\pi^2} \left\{ C_W g^2 + C_H \lambda - \sum_f C_f y_f^2 \right\} \Lambda^2_{\text{UV}} = -\lambda(246 \text{ GeV})^2. \quad (2.4)$$

The "one-loop improved" relation (2.4) essentially depends on all of SM free parameters. (The $SU(3)$ coupling g_s enters at higher loop orders which are not shown here.) Hence, even though cancellations are technically possible (by adjusting the tree-level mass m^2_{tree} or λ) and despite the fact that in the presence of a tree-level mass one can simply subtract the infinity by introducing an appropriate counter term, such procedures do not seem reasonable unless $\Lambda_{\text{UV}} \sim \mathcal{O}(\nu)$. For example, for a cut-off of the order of magnitude of the Planck mass $\Lambda_{\text{UV}} \sim M_{\text{Planck}} \sim 10^{19}$ GeV one needs to fine-tune the value of $\nu = 246$ GeV by 10^{17} orders of magnitude (and ν^2, which appears in eq. (2.4), by 10^{34} orders of magnitude)!

This naturalness problem is often referred to as the hierarchy problem: Naively, one expects that electroweak symmetry breaking would have occurred near the Planck scale $M_{\text{Planck}} \sim 10^{19}$ GeV and not at the electroweak scale $M_{\text{Weak}} \sim g\nu \sim 10^2$ GeV (which is measured with astonishing precision, for example, see M_Z and M_W measurements given in the previous chapter). Since the SM does not contain the means to address this issue, it follows that the resolution of the hierarchy problem and any understanding of the specific choice of the electroweak scale must lie outside the SM. To reiterate, the instability of the infrared SM Higgs potential implies that the model cannot be fundamental while it severely constrains the possible extrapolations and embeddings of the SM.

Fig. 2.1. The gauge interaction contributions to the quadratic divergence.

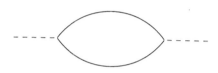

Fig. 2.2. The Yukawa interaction contribution to the quadratic divergence.

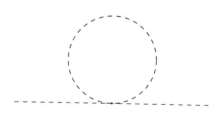

Fig. 2.3. The scalar self-interaction contribution to the quadratic divergence.

2.2 Possible Avenues

Once it has been argued that the SM cannot be a fundamental theory, the obvious question arises: What are the possible avenues that could be pursued in order to consistently extend it. Our discussion of the hierarchy problem clearly indicates two possibilities: (i) The existence of a fundamental scale near the electroweak scale corresponding to new gauge (or gravitational) dynamcis which effectively serves as a cut-off on any extrapolation of the SM and its parameters. Such dynamics presumably render one or more elements (e.g. a fundamental Higgs field, two hierarchically distinctive scales, perturbative validity) of the hierarchy problem obsolete and the corrections eqs. (2.1)–(2.3) are cut off by its low scale and are not dangerous. (ii) A theory containing a fundamental Higgs field and which extends and extrapolates the SM within perturbation theory (and in four dimensions). Such a theory contains all the elements of the hierarchy problem yet corrections of the form of eqs. (2.1)–(2.3) "miraculously" vanish. The essence of the theory in this case is in naturally explaining the "miracle". Both options imply the prediction of a new field content at or near the electroaweak scale and which is associated with either the nearby more fundamental "cut-off" scale (let it be resonances associated with nearby strong interactions or states associated with the compactification of an extended space-time geometry near the electroweak scale), or alternatively, with the cancellation of (a larger more complete set of) quadratically divergent corrections in a perturbative theory. This realization is a main driving force, side by side with the search for the Higgs boson itself, in current studies of particle physics at electroweak energies and beyond, and provides the basis for this manuscript.

Various proposals exist that either postulate $\Lambda_{\rm UV} \sim 4\pi\nu$ or otherwise provide for a natural cancellation of divergences. We will sketch the various ideas which underlie those frameworks here, and will further elaborate on the supersymmetry framework in the next section.

2.2.1 Strong Interactions

Certain frameworks postulate that the Higgs field is a meson, a composite of fermions, rather than a fundamental field. In this case the notion of a cut-off scale and of a confinement scale are entangled and, in fact, the solution to the hierarchy problem is the elimination of scalar bosons from the fundamental theory. It duplicates (and at much higher energies) in its philosophy the QCD picture in which the low-energy degrees of freedom are mesons and baryons while at high energy these are the quarks and gluons. Under this category fall the different versions of technicolor models, top-color models, and their various offsprings. (For a recent review, see Refs. [9, 10, 11].)

The models predict not only additional matter, but also strong and rich dynamics "nearby" (not far from current experimental energies). Strong dynamics, in general, does not tend to completely decouple from its nearby low-energy limit (the weak scale, in this case), leading, in most cases, to many predictions at observable energies. For example, the models typically predict new FCNC as well as (oblique) quantum corrections to the W and Z masses and interactions. (As mentioned in the previous chapter, the latter lead to calculable shifts from SM tree-level relations such as $M_W = M_Z \cos\theta_W$, which is measured by the ρ-parameter.) While a detailed critique can be found in Altarelli's lectures [3], no corresponding deviations from SM predictions (when including its own quantum corrections to such observables) were found, seriously undermining the case for such an embedding of the SM.

Attempts to solve this contradiction are often based on distinguishing, for example, the fermion-mass generation sector from the sector responsible to electroweak symmetry breaking, leading to complex multi-layer constructions. It is also interesting to note that strong dynamics typically (counterexamples, however, exist [12]) predict for the Higgs boson mass $m_\eta \gg \nu$ (consider the pions in QCD), contrary to current implications based on fits to electroweak data [13] $m_\eta \lesssim \nu$. (Such fits are based, however, on the assumption that the new physics decouples from many of the SM observables. A heavier Higgs boson is allowed otherwise [14, 15].) One should bare in mind, however, that by its nature non-perturbative strong dynamics do not always allow for reliable calculations of SM quantities (consider the Postmodern Technicolor [16] model, for example), and it is therefore difficult to rule it out.

2.2.2 Extra Spatial Dimensions

A different and more recent proposal is of a framework with a low-energy cut-off and (possibly) a fundamental Higgs field in the background of an extended space-time geometry (and more then three spatial dimensions). Theories with extra low-energy (very large) compacitified dimensions where first argued to be consistent with all data (for two or more extra dimensions) by Arkani-Hamed, Dimopoulos and Dvali [17]. In short, it is postulated that our four-dimensional universe resides on a four-dimensional submanifold of an extended space-time. Such theories must contain towers of Kaluza-Klein states which parameterize the excitations of those states propagating in the extra dimensions and which contain $M \sim \mathcal{O}(\text{TeV})$ or heavier states. Such towers may include the graviton (and other fields describing quantum gravity such as various moduli) but could also include for not "so low-energy dimensions" (*i.e.*, for extra dimensions smaller than originally proposed in Ref. [17]) towers of the usual and extended gauge fields and/or the usual and extended matter and Higgs fields, leading to a large variety of possibilities and constraints on the size and number compactified dimensions $R \gg (M_{\text{Planck}}^{4d})^{-1}$.

The relation between the four-dimensional to $4 + n$-dimensional Planck constants reads

$$M_{\text{Planck}}^{4d} = (M_{\text{Planck}}^{(4+n)d})^{\frac{n+2}{2}} R^{\frac{n}{2}}, \tag{2.5}$$

relating the size and number of compactified dimensions. Relation (2.5) explains the weakness of gravity as we know it in terms of extra large (but with particular size) dimensions rather than in terms of a large Planck mass. $R^n \equiv V$ is the volume of the compactified n-dimensional space (and R^{-1} is the compactification scale).

In this framework, the Higgs boson is usually assumed to be a fundamental particle, but this need not be the case. Rather, it depends on the assumed details of the "bulk" physics (where the "bulk" refers to the the the extra-dimension volume V).

The theoretical limit of a small volume V ($R^{-1} \sim M_{\mathrm{Planck}}^{4d}$) is also very interesting. It occurs in what is often called M-theory [18, 19] (which is related to the discussion of unification in Chap. 6). It also appears in an alternative realization of these ideas which is based on the notion that space-time geometry is modified such that the full space-time metric is not factorizable (that is, it cannot be written as a direct sum of the metric on the four-dimensional submanifold and of the metric in the extra dimensions, as is implicitly assumed in deriving eq. (2.5)). Rather, the four-dimensional metric is assumed to be multiplied by an exponential function of the additional dimension(s), the warp factor. It was shown by Randall and Sundrum [20] that in the case of two four-dimensional manifolds, or "walls" (one of which corresponds to our world, while gravity is localized on the other "wall"), in an anti-de Sitter background, and which are separated by one extra dimension, it is sufficient to stabilize the size of the compactified dimension R at a small (yet particular) value in order to realize the hierarchy as seen in our four-dimensional world: The hierarchy is given by $\exp(-2MR\pi)$, where M here is some fundamental scale in the full theory (and is assumed to be of the order of M_{Planck}). A rescaling of the coordinates by the square root of the inverse of the warp factor $\exp(MR\pi)$ leads to a similar picture but now the two scales play the reverse roles with the fundamental Planck scale being of the order of the electroweak scale, in the spirit of the original proposal of Ref. [17].

The hierarchy problem is not eliminated in the modified geometry (extra-dimensional) frameworks but is only rephrased in a different language. The large extra dimensions must be stabilized so that the volume V is appropriately fixed. The stabilization of the extra dimensions at particular values is the hierarchy problem of models with extra large (or equivalently, low-energy) dimensions. Intuitively, one expects its solution within a framework of quantum gravity. On the other hand, the only candidate to describe quantum gravity, namely string theory, does not provide the answer, given current knowledge. (It was shown [21], however, that in the Randall-Sundrum [20] setup one could consistently take the limit of a non-compactified infinitely-large extra dimension $R \to \infty$ while having gravity localized on the remaining four-dimensional "wall".) Furthermore, if the $d + n$ Planck mass (or the "warped" quantities, in the alternative picture) are not sufficiently close to the weak-scale then the usual hierarchy problem re-appears.

Many issues have been recently discussed in the context of extra dimensions, including the unification of couplings (first discussed in Ref. [22]), proton decay (see, for example, Ref. [23]), *etc*. The discussions are far from conclusive due to the lack of a concrete framework on the one hand, and strong dependence on the details of the full theory, *i.e.*, boundary conditions, on the other hand. (Note in particular that the issue of possible new contributions to FCNC is difficult to address in the absence of concrete (and complete) realizations of these ideas.) Even though it is intriguing that low-energy extra dimensions may still be consistent with all data, significant constraints were derived in the literature in the recent couple of years and no indications for such a dramatic scheme exist in current data. These ideas may be further tested in collider experiments. A generic prediction in many of the extra-dimension schemes is of large effective modifications of the SM gauge interactions (due, for example, to graviton emission into the bulk of the extended volume) which can be tested in the next generation of hadron and lepton colliders, if the $(4 + n)d$ Planck mass is not too large. Other effects of low-energy gravity may include a spin 2 graviton resonance [24] and pure gravitational effects (one may even speculate on balck-hole[3] production at colliders), if gravity is indeed sufficiently strong at sufficiently low energies.

2.2.3 Conformal Theories

On a different note, it was also argued (but not demonstrated) that one could construct theories which are conformally (scale) invariant above the weak scale and hence contain no other scales, no dimensionful parameters, and trivially, no hierarchy problem [26]. This proposal combines the notion of a fundamental weak-scale cut-off (above which nature is truly scale invariant) with that of a cancellation of all divergences (due to the conformal symmetry). Again, this is only a rephrasing of the problem since conformal invariance must be broken and by the right amount. Furthermore, this approach faces severe and fundamental difficulties [27] in constructing valid examples.

2.2.4 Supersymmetry

Unlike in all of the above schemes, one could maintain perturbativity to an arbitrary high-energy scale if dangerous quantum corrections are canceled to all orders in perturbation theory (a cancellation which does not require conformal invariance discussed in the previous section). Pursuing this alternative, one arrives at the notion of weak-scale supersymmetry. Supersymmetry ensures the desired cancellations by relating bosonic and fermionic **degrees of freedom** (e.g. for any chiral fermion the spectrum contains a complex scalar field, the s-fermion) and **couplings** (e.g. quartic couplings depend on gauge and Yukawa couplings). This is done by simply organizing fermions and bosons in "spin multiplets" (so that chirality is attached to the bosons by their association with the fermions). Henceforth, supersymmetry is able to "capitalize" on the sign difference between bosonic and fermionic loops, which is insignificant as it stands in the SM result (2.4). This is the avenue

---

[3] Black hole were also argued [25] to induce dangerous low-energy operators in these theories.

that will be pursued in these notes, and the elimination of the dangerous corrections will be shown below explicitly in a simple example. A detailed discussion will follow in the succeeding chapters. We conclude this chapter by elaborating on the status of (weak-scale) supersymmetry, which further motivates our choice.

2.2.5 Mix and Match

We note in passing that it is possible to link together various avenues. Theories with extra dimensions may very well contain weak-scale supersymmetry, depending on the relation between the (string) compactification scale and the supersymmetry-breaking scale [28]. In fact, it is difficult to believe that the stabilization of the large extra dimensions does not involve supersymmetry in some form or another. It was also suggested that supersymmetric QCD-like theories may play some role in defining the low-energy theory (for example, that some fermions and their corresponding sfermions are composites [29, 30].) Of course, strong dynamics may reside in extra dimensions (for a recent proposal, see Ref. [31]) and all three avenues may be realized simultaneously. These possibilities lie beyond the scope of this manuscript.

2.3 More on Supersymmetry

Aside from its elegant and nearly effortless solution to the hierarchy problem (which will occupy us in the next part), supersymmetric models suggest that electroweak symmetry is broken in only a slightly "stronger" fashion (*i.e.*, two fundamental Higgs doublets acquire a *vev*) than predicted in the SM (where only one Higgs doublet is present). Furthermore, in many cases the more complicated Higgs sector is reduced to and imitates the SM situation with the additional theoretical constraint $m_\eta \lesssim 130 - 180$ GeV. This is a crucial element in testing the supersymmetry framework and we will pay special attention to this issue in Part III. In fact, one typically finds that the theory is effectively described by a direct sum of the SM sector (including the SM-like Higgs boson) and of a superparticle (sparticle) sector, where the latter roughly decouples (for most practical purposes) from electroweak symmetry breaking (*i.e.*, its spectrum is to a very good approximation $SU(2)_L \times U(1)_Y$ conserving and is nearly independent of electroweak symmetry breaking *vev's*). The sparticle mass scale provides the desired cut-off scale above which the presence of the sparticles and the restoration of supersymmetric relations among couplings ensure cancellation of quadratically divergent quantum corrections. Hence, the super-particle mass scale is predicted to be $M_{\mathrm{SUSY}} \sim \mathcal{O}(100\,\mathrm{GeV} - 1\,\mathrm{TeV})$ with details varying among models. The mass scale M_{SUSY} is also a measure, in a sense, of supersymmetry breaking in nature (e.g. the obvious observation that a fermion and the corresponding sfermion boson are not degenerate in mass). Its size implies that supersymmetry survives (and describes the effective theory) down to the weak scale. More correctly, it is M_{SUSY} that defines in this picture the weak scale $\nu \sim M_{\text{SUSY}}$ (up to a possible loop factor and dimensionless couplings). These issues will be discussed in some detail in the following chapters.

The electroweak symmetry conserving structure of the sparticle spectrum sharply distinguishes the models from, e.g. technicolor models. It implies that the heavy sparticles do not contribute significantly even at the quantum level to electroweak (or more correctly, to the related custodial $SU(2)$ symmetry) breaking effects such as the ρ-parameter. (This is generically true if all relevant superparticle masses are above the 200-300 GeV mark.) This is consistent with the data, which is in nearly perfect agreement with SM predictions. Returning to the Higgs mass, it is particularly interesting to note that the same analysis of various electroweak observables (*i.e.*, various cross-sections, partial widths, left-right and forward-backward asymmetries, etc., including their SM quantum corrections) which determines the value of the ρ-parameter, suggests that the SM Higgs boson (or the SM-like Higgs boson $h^0 \simeq \eta$ in supersymmetry) is lighter than 200-300 GeV [13], which is consistent with the predicted range in supersymmetry, and more generally, with the notion of a fundamental Higgs particle (rather that a composite Higgs with a mass $m_\eta \sim \Lambda_{\text{UV}}$ which is a factor of $2-4$ heavier). Both the data and supersymmetry seem to suggest

$$g_{\text{Weak}}\nu \lesssim m_\eta \lesssim \nu, \tag{2.6}$$

where $g_{\text{Weak}}\nu \equiv M_Z$. The lower experimental bound of $\mathcal{O}(100)$ GeV is from direct searches. (The origin of the theoretical bounds will be explored at a later point).

It should be stressed that one is able to reach such strong conclusions regarding the Higgs mass primarily because the t-quark (top) mass [7] $m_t = 174 \pm 5$ GeV is measured and well known, hence, reducing the number of unknown parameters that appear in the expressions of the quantum mechanically corrected SM predictions for the different observables. Indeed, it was m_t which appears quadratically in such expressions (rather than only logarithmically as the Higgs mass) that was initially determined by a similar analysis long before the t-quark was discovered and its mass measured [32] in 1995. (For example, Langacker and Luo found [33] $m_t = 124 \pm 37$ GeV in 1991.) The most important lesson learned from the early analysis was that the top is heavy, $m_t \sim \nu$ (in analogy to the current lesson that the Higgs boson is light, with light and heavy assuming a different meaning in each case). Once we discuss in Part III the renormalization of supersymmetric models and the dynamical realization of electroweak symmetry breaking that follows, we will find that a crucial element in this realization is a sufficiently large t-Yukawa coupling, *i.e.*, a sufficiently large $m_t \simeq y_t \nu$. Indeed, this observation corresponds to the successful prediction of a heavy top in supersymmetry (though there are small number of cases in which this requirement can be evaded).

A somewhat similar but more speculative issue has to do with the extrapolation to high energy of the SM gauge couplings. Once loops due to virtual (*i*) sparticles and (*ii*) the second Higgs doublet are all included in the extrapolation at their predicted mass scale of $\mathcal{O}(100\,\text{GeV} - 1\,\text{TeV})$, then one finds that the couplings unify two orders of magnitude below the Planck scale and with a perturbative value (provided that the spectrum contains exactly two Higgs doublets and their super-partners). This enables the consistent discussion of grand-unified and string theories, which claim such a unification, in this framework. (Note that supersymmetry is a natural consequence of (super)string theory, though its survival to low energies is an independent conjecture.) This result was known qualitatively for a long time, but it was shown more recently to hold once the increasing precision in the measurements of the gauge coupling at the weak scale is taken into account, and it re-focused the attention of many on low-energy supersymmetry. (As we discuss in Chap. 11, the large unification scale may play an important role in the smallness of neutrino masses.)

On the other hand, no known extension of the SM can imitate the simplicity by which FCNC and LFV are suppressed. Supersymmetry is no exception and like in the case of any other low-energy extension, new potentially large contributions to FCNC (and in some cases to LFV) could arise. Their suppression to acceptable levels is definitely possible, but strongly restricts the "model space". The absence of large low-energy FCNC is the most important information extracted from the data so far with regard to supersymmetry, and as we shall see, one's choice of how to satisfy the FCNC constraints (the "flavor problem") defines to great extent the model.

In conclusion, once large contributions to FCNC are absent, the whole framework is consistent with all data, and furthermore, it is successful with regard to certain indirect probes. Its ultimate test, however, lies in direct searches in existing and future collider experiments. (See Zeppenfeld [34] for a discussion of collider searches and Halzen [35] for the discussion of possible non-collider searches.) In the course of these notes all of the above statements and claims regarding supersymmetry will be discussed in some detail and presented using the tools and notions that we are about to define and develop.

Part II

Supersymmetry Bottom-Up:
Basics

3. Bottom-Up Construction

Though supersymmetry corresponds to a self-consistent field-theoretical framework, one can arrive at its crucial elements by simply requiring a theory which is at most logarithmically divergent: Logarithmic divergences correspond to only a weak dependence on the ultraviolate cut-off Λ_{UV} and can be consistently absorbed in tree-level quantities, and thus understood as simply scaling (or renormalization) of the theory. Unlike quadratic divergences – which imply strong dependence on the ultraviolate cut-off and therefore need to be fine-tuned away, logarithmic divergences do not destabilize the infrared theory but only point towards its cut-off scale.

As we shall see, the mere elimination of the quadratic divergences from the theory renders it "supersymmetric", and the importance of supersymmetry to the discussion of any consistent high-energy extension of the SM naturally follows. This bottom-up construction of "supersymmetry" is an instructive exercise. The simple example of a (conserved) $U(1)$ gauge theory will suffice here and will be considered below, followed by the consideration of a Yukawa (singlet) theory. This exercise could be repeated, however, for the case of a (conserved) non-Abelian gauge theory and in the case of the SM (with a spontaneously broken non-Abelian gauge symmetry and non-singlet Yukawa couplings) which are technically more evolved. Suggested readings for this section include lecture notes Refs. [36, 37, 38].

3.1 Cancellation of Quadratic Divergences in a $U(1)$ Gauge Theory

We set to build a theory which contains at least one complex scalar (Higgs) field ϕ^+ (with a $U(1)$ charge of $+1$) and which is at most logarithmically divergent. In particular, we would like to extend the field content and fix the couplings such that all quadratic divergences are eliminated. For simplicity, assume a massless theory and zero external momenta $p^2 = 0$. All terms which are consistent with the symmetry are allowed. (The Feynamn gauge, with the gauge boson propagator given by $D_{\mu\nu} = -ig_{\mu\nu}/q^2$, is employed below.)

As argued above, the most obvious "trouble spot" is the scalar two-point function (Figs. 2.1–2.3). A $U(1)$ theory with only scalar (Higgs) field(s) receives at one-loop positive contributions $\propto \Lambda_{\mathrm{UV}}^2$ from the propagation of the gauge and Higgs bosons in the loops (Figs. 2.1 and 2.3). They are readily evaluated in the $U(1)$ case to read

$$\delta m_\phi^2|_{\text{boson}} = + \left\{ (4-1)g^2 + \lambda \frac{N}{2} \right\} \int \frac{d^4q}{(2\pi)^4} \frac{1}{q^2}, \qquad (3.1)$$

where g and λ are the gauge and quartic couplings, respectively, and $N = 8$ is a combinatorial factor ($N = 2$ for non-identical external and internal fields). Keeping only leading terms, the integral is replaced by $\Lambda_{\text{UV}}^2/16\pi^2$.

Obviously, there must be a negative contribution to δm_ϕ^2, if cancellation is to be achieved. A fermion loop equivalent to Fig. 2.2, which could cancel the contribution from (3.1), implies a new Yukawa interaction. In order to enable such an interaction and the subsequent cancellation, we are forced to introduce new fields to the theory: (i) A chiral fermion ψ_V with gauge quantum numbers identical to those of the gauge boson ($i.e.$, neutral). (It follows that ψ_V is a real Majorana fermion.) (ii) A chiral fermion ψ_ϕ^+ with gauge quantum numbers identical to those of the Higgs boson (i.e., with $U(1)$ charge of $+1$). Given this fermion spectrum, the Lagrangian can now be extended to include the Yukawa interaction $y\phi^+\bar{\psi}_\phi\psi_V$ where y is a new (Yukawa) coupling. Note the equal number (two) of bosonic and fermionic degrees of freedom ($d.o.f.$) with the same quantum numbers in the resulting spectrum. In particular, the requirement of identical gauge quantum number assignments guarantees the gauge invariance of the Yukawa interaction. We will adopt the equality of fermion and boson number of $d.o.f.$ as a guiding principle.

Let us redefine $\lambda \equiv bg^2$ and $y \equiv cg$. One arrives at

$$\delta m_\phi^2|_{\text{fermion}} = -2c^2g^2 \int \frac{d^4q}{(2\pi)^4} \frac{1}{q^2}. \qquad (3.2)$$

The coefficients b and c must be predetermined by some principle or otherwise a cancellation is trivial and would simply correspond to fine-tuning (of the coefficients, this time). The principle is scale invariance. Even though the determination of the proportionality coefficient cannot be done unambiguously at the level of our discussion so far, it follows from the consideration of the logarithmic divergences and the requirement that any proportionality relation is scale invariant (i.e., that it is preserved under renormalization group evolution: $db/d\ln\Lambda = dc/d\ln\Lambda = 0$). However, before doing so we note an inconsistency in our construction above which must be addressed in order to facilitate the discussion of scale invariance.

Once fermions are introduced into the $U(1)$ gauge theory, one has to ensure that the (massless) fermion content is such that the theory is anomaly free, that is $\text{Tr}Q_{\psi_i} = 0$ (where the trace is taken over the $U(1)$ charges of the fermions). The gauge fermion ψ_V carries no charge (it carries charge in non-Abelian theories, but being in the traceless adjoint representation it still does not contribute) so only the introduction of one additional chiral fermion ψ_ϕ^- with a $U(1)$ charge of -1 is required in order to render the theory anomaly free. The additional fermion ψ_ϕ^- itself does not lead to new one-loop quadratically divergent contribution to the ϕ^+ two-point function. However, our newly adopted principle of equal number of bosonic and fermionic d.o.f. suggests that it should be accompanied by a complex scalar Higgs boson ϕ^-, which also carries a $U(1)$ charge of -1, and it leads to new divergences. The most general gauge invariant quartic potential is now that of a two Higgs model,

$$
\begin{aligned}
V_{\phi^4} = &+\lambda_1 |\phi^+|^2 |\phi^+|^2 + \lambda_2 |\phi^-|^2 |\phi^-|^2 + \lambda_3 |\phi^+|^2 |\phi^-|^2 + \lambda_4 |\phi^+ \phi^-|^2 \\
&+ \left\{ \lambda_5 (\phi^+ \phi^-)^2 + \lambda_6 |\phi^+|^2 \phi^+ \phi^- + \lambda_7 |\phi^-|^2 \phi^+ \phi^- + h.c. \right\},
\end{aligned}
\tag{3.3}
$$

where we adopted standard notation for the couplings, replacing $\lambda \to \lambda_1$. In the following we will define $\tilde{\lambda}_3 = \lambda_3 + \lambda_4 \equiv \tilde{b} g^2$.

With the introduction of ψ^- the theory is now consistent, and with the additional introduction of ϕ^- it is possible to require scale invariant relations between the different coefficients. Doing so one finds a unique solution $b = 1/2$, $\tilde{b} = -1$, and $c = \sqrt{2}$ (where λ_1 and $\tilde{\lambda}_3$ one-loop mixing is accounted for). It is straightforward to show that the same arguments lead to $\lambda_2 = \lambda_1$. Similarly, $\lambda_{6,7}$ (λ_5) may be fixed by consideration of one-loop (two-loop) wave-function mixing between ϕ^+ and ϕ^-, and must vanish.

Thus, we finally arrive at the desired result: A vanishing total contribution

$$
\delta m_\phi^2 |_{\text{total}} = g^2 \left\{ 3 - 2c^2 + 4b + \tilde{b} \right\} \int \frac{d^4 q}{(2\pi)^4} \frac{1}{q^2} = 0,
\tag{3.4}
$$

where the last term is from ϕ^- circulating in the loop (with $N = 2$). One concludes that in a consistent theory with (i) equal number of bosonic and fermionic d.o.f., and (ii) scale-invariant relations among the respective couplings, the scalar two-point function is at most logarithmically divergent! The combination of these two conditions defines "supersymmetry".

This is a remarkable result that suggests more generally the interaction terms, rewritten in a more compact form,

$$
\mathcal{L}_{\text{int}} = \sqrt{2} g \sum_i Q_i \left(\phi_i \bar{\psi}_i \psi_V + h.c. \right) - \frac{g^2}{2} \left\{ \sum_i Q_i |\phi_i|^2 \right\}^2,
\tag{3.5}
$$

where we denoted explicitly the dependence on the $U(1)$ charge Q_i. (This dependence is apparent above in $b = Q_{\phi^+}^2/2$ and $\tilde{b} = 2Q_{\phi^+}Q_{\phi^-}/2$.) The most important lesson is that the theory contains only one free coupling: The gauge coupling. This could have been anticipated from our introductory discussion of divergences, and further suggests that the Lagrangian (3.5) is derived in a supersymmtric theory from the gauge Lagrangian. (This will be illustrated in the following chapter.)

Our discussion so far dealt with the elimination of the $\mathcal{O}(\Lambda_{\text{UV}}^2)$ terms at one-loop order. They re-appear at higher loop orders, for example, two-loop

$\mathcal{O}((g^2/16\pi^2)^2 \Lambda_{\rm UV}^2)$ contributions to the two-point function, etc. It is straightforward, though technically evolved, to show that the conditions found above are not only sufficient for the elimination of the undesired divergences at one-loop order, but rather suffice for their elimination order by order. Thus, "supersymmetry", which postulates equal number of bosonic and fermionic d.o.f. and specific scale-invariant relations among the various couplings, is indeed free of quadratic divergences at all orders in perturbation theory. This is a consequence of the non-renormalization theorems [1] of supersymmetry.

3.2 Cancellation of Quadratic Divergences in a Yukawa Theory

The above prescription extends also to non-gauge interactions. For example, consider a theory with a gauge-singlet complex scalar and a gauge-singlet chiral fermion fields, ϕ_s and ψ_s, respectively (or in a more compact terminology, a superfield S with a scalar and a chiral fermion components ϕ_s and ψ_s, respectively): The theory contains an equal number of fermionic and bosonic d.o.f. The interaction Lagrangian

$$\mathcal{L}_{\rm int} = y\phi_s\psi_s\bar{\psi}_s + A\phi_s^3 + C\phi_s^*\phi_s^2 + h.c. - \lambda(|\phi_s|^2)^2 \tag{3.6}$$

leads to quadratically divergent contributions to the scalar two-point function (from the equivalents of diagrams 2.3 and 2.2),

$$\delta m_{\phi_s}^2|_{\rm total} = 4(\lambda - y^2) \int \frac{d^4q}{(2\pi)^4} \frac{1}{q^2}. \tag{3.7}$$

(Note the additional factor of 2 in the fermion loop from contraction of Weyl indices.) The cancellation of (3.7) leads to the relation $\lambda = y^2$ among the dimensionless couplings. Again we find the same lesson of constrained quartic couplings!

The singlet theory contains another potential quadratic divergence, in the scalar one-point function (the tadpole). The contribution now depends on the dimensionful trilinear coupling C. (The symbol A' is often used instead of C). Its dimensionality prevents it from contributing to (3.7), but if non-vanishing it will contribute to the quadratically divergent tadpole. Its contribution may be offset by a fermion loop, but only if the fermion is a massive (Dirac) fermion with (a Dirac) mass $\mu \neq 0$ so that its propagator contains a term $\propto \mu/q^2$ (which, by dimensionality, is the relevant term here). Both divergent loop contributions to the singlet one-loop function are illustrated in Fig. 3.1. Evaluation of the loop integral gives for the leading contributions to the tadpole

$$\delta T_{\phi_s} = \{y{\rm Tr}I\mu - 4C\} \int \frac{d^4q}{(2\pi)^4} \frac{1}{q^2}. \tag{3.8}$$

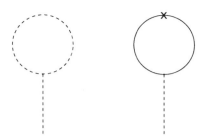

Fig. 3.1. The divergent contributions to the one-point function of the singlet.

The trace over the identity is four, leading to the condition $C = y\mu$ so that the tadpole is canceled (at leading order).

As in the gauge theory case, the relevant relations, $\lambda = y^2$ and $C = y\mu$, can be shown to be scale-invariant, and to lead to cancellation of quadratic divergences order by order in perturbation theory. These relations extend to non-singlet Yukawa theories such as the SM.

Note that the trilinear interaction $A\phi_s^3$ as well as a mass term $m^2|\phi_s|^2$, do not lead to any quadratic divergences nor upset their cancellations as described above. Yet, one defines the supersymmetric limit as $m^2 = \mu^2$ and $A = 0$, corresponding to a fermion-boson mass degeneracy.

The singlet Yukawa theory described above is essentially the Wess-Zumino model [39], which is widely regarded as the discovery of supersymmetry, and from which stems the field-theoretical formulation of supersymmetry.

Exercises

3.1 Using the Feynamn rules of a $U(1)$ (QED) theory, confirm the numerical factors in eqs. (3.1) and (3.2).

3.2 Repeat our exercise for ϕ^- and show $\lambda_2 = \lambda_1$.

3.3 Repeat the $U(1)$ example in the Landau gauge, $D_{\mu\nu} = -\frac{i}{q^2}\left\{g_{\mu\nu} - \frac{q_\mu q_\nu}{q^2}\right\}$.

3.4 Show that $\lambda_{6,7} = 0$ by considering the loop corrections to a $m_{+-}^2 \phi^+ \phi^-$ mass term. Repeat the exercise for λ_5 which is constrained by two-loop corrections.

3.5 Show that a term $A\phi_s^3$ does not lead to any new quadratic divergences.

3.6 Repeat the relevant loop calculations with a scalar mass $m^2 \neq \mu^2$ (boson fermion mass non-degeneracy) and show (for example, by expanding the scalar propagator) that it does not lead to any new quadratic divergences (for a finite m^2) but only to finite corrections $\propto m^2 - \mu^2$.

3.7 Above, we implicitly assumed a unique cut-off scale on both fermionic and bosonic loops. Assume that $\Lambda_{\mathrm{UV}}^{boson} = r\Lambda_{\mathrm{UV}}^{fermion}$ for some r. Derive the relations between the relevant couplings in the last example as a function of r. For what values of r are the relations scale (*i.e.*, renormalization-group) invariant?

4. Supersymmetry

In the previous chapter it was shown that in trying to eliminate quadratic divergences in two- and one-point functions of a scalar field one is led to postulate an equal number of bosonic and fermionic $d.o.f.$, as well as specific relations among the couplings in the theory. (Uniqueness was achieved once the relations were required to be scale invariant.) From the consideration of the scalar-vector interaction one is led to postulate new gauge-Yukawa and gauge-quartic interactions. From the consideration of a (non-gauge) Yukawa interaction one is able to fix the scalar trilinear and quartic couplings. These relations suggest a small number of fundamental couplings, and hence a small number of fundamental (super)fields: Fermions and bosons may be arranged in spin multiplets, the superfields, which interact with each other. In this (supersymmetric) language the Lagrangian will contain a small number of terms. Nevertheless, when rewriting the interactions and kinetic terms in terms of the superfield components – the individual boson and fermion fields – the Lagrangian will contain a large number of terms which exhibit what may naively seem to be mysterious correlations.

The field theoretical formalism that realizes this elegant idea is super-symmetry, with the spin multiplets as its fundamental building blocks which are subject to the supersymmetric transformations: Notions which bring to mind low-energy flavor symmetries and their fundamental building blocks, the isospin multiplets. Isospin transformations act on the proton-neutron doublet. supersymmetric "spin" transformations act on the boson-fermion multiplet, the superfield. In order to define supersymmetry as a consistent (at this point, global) symmetry one needs to extend the usual space-time to superspace by the introduction of (not surprisingly equal number of) "fermionic" (or spino-rial) coordinates

$$\{x_\mu\}_{\mu=1,\ldots,4} \to \left\{ x_\mu, \begin{pmatrix} \theta_\alpha \\ \bar{\theta}_{\dot{\beta}} \end{pmatrix} \right\} \begin{matrix} \mu = 1,\ldots,4 \\ \alpha, \dot{\beta} = 1, 2. \end{matrix}$$

The new coordinate θ is a spin $1/2$ object, $i.e.$, a Grassmann variable which carries two $d.o.f.$ (and which, as shown below, carries mass dimensions).

The formalism of supersymmetric field theories can be found in the literature: Refs. [40, 41, 42, 43, 44, 37, 45] provide a useful sample of such reviews, while Wess and Bagger [46] provide a complete, though somewhat condensed, discussion. It is not our purpose to introduce supersymmetry in an elaborate way, but rather to outline it for completeness and to illustrate some of its more intuitive aspects (which are also the more relevant ones for the discussion in the rest of these notes). We begin with a brief introduction to Grassmann algebra. Next, we will define a chiral superfield and use it to illustrate the supersymmetry algebra. We then will fill in some of the details such as kinetic and interaction terms. We conclude with brief discussions of a gauge theory and, finally, of gauging supersymmetry itself.

4.1 Grassmann Algebra

The Grassmann variables are anticommuting variables,

$$\{\theta_{i\alpha}, \theta_{j\beta}\} = 0 \tag{4.1}$$

with the antisymmetric tensor

$$\epsilon_{\alpha\beta} = \begin{pmatrix} 0 & 1 \\ -1 & 0 \end{pmatrix}$$

as metric. One has, for example, $\theta_{i\alpha} = \epsilon_{\alpha\beta}\theta_i^\beta$, where Greek indices correspond to spinorial indices while Latin indices (which are sometimes omitted) label the variable. It is customary to reserve a special notation to the hermitian conjugate (as above) $(\theta_i^\alpha)^\dagger = \bar{\theta}_i^{\dot\alpha}$. The Pauli matrices, which act on Grassmann variables, are then written as $\sigma^\mu_{\alpha\dot\alpha}$.

One defines derivatives and integration with respect to the Grassmann variables as

$$\frac{d\theta_{i\alpha}}{d\theta_{j\beta}} = \delta_i^j \delta_\alpha^\beta, \tag{4.2}$$

$$\int d\theta_i = 0; \quad \int d\theta_{i\alpha}\theta_{j\beta} = \delta_{ij}\delta_{\alpha\beta}. \tag{4.3}$$

In eq. (4.3) the first relation follows from the invariance requirement $\int d\theta_i f(\theta_i) = \int d\theta_i f(\theta_i + \theta_{i_0})$ where f is an arbitrary function, e.g. $f(\theta_i) = \theta_i$. The second relation is simply a choice of normalization. The derivative operator is often denoted with the short-hand notation ∂_{θ_j}. Note that derivation (4.2) and integration (4.3) of Grassmann variables are equivalent operations.

The function $f(\theta_i)$ can be expanded as a polynomial in θ_i. However, the anticommuting nature of θ_i,

$$\theta_i^\alpha \theta_{i\alpha} \theta_{i\alpha} = \tfrac{1}{2}\theta_i^\alpha \{\theta_{i\alpha}, \theta_{i\alpha}\} = 0,$$

implies that $f(\theta_i)$ can at most be a quadratic polynomial in θ_i. In particular, any expansion is finite! Using its polynomial form and the above definitions, the integration-derivation equivalence follows for any function of Gramssmann variables,

$$\int d\theta_i f(\theta_i) = \frac{df(\theta_i)}{d\theta_i}. \tag{4.4}$$

Another useful relation that will be implicitly used in some of the expansions below is

$$\theta_i^\alpha \theta_i^\beta = -\frac{1}{2}\epsilon^{\alpha\beta}\theta_i^\alpha \theta_{i\alpha}. \tag{4.5}$$

In the following, we will omit Latin indices when discussing an expansion in only one Grassmann variable.

4.2 The Chiral Superfield: The L-Representation

Equipped with the algebraic tools, our obvious task is to embed the usual bosons and fermions into spin multiplets, the superfields. Begin by considering a Higgs field, $i.e.$, a (complex) scalar field $\phi(x_\mu)$. Requiring the corresponding superfield $\Phi(x_\mu, \theta_\alpha, \bar\theta^\beta)$ – which is a function defined in superspace rather than only in space-time – to carry the same quantum numbers and mass dimension as ϕ, and expanding it, for example, in powers of θ_α (with x_μ-dependent coefficients), one arrives at the following embedding:

$$\Phi(x_\mu, \theta_\alpha) = \phi(x_\mu) + \theta^\alpha \psi_\alpha(x_\mu) + \theta^\alpha \theta_\alpha F(x_\mu). \tag{4.6}$$

(We will return to the issue of an expansion in $\bar\theta^\beta$ below.) From the spinorial nature of θ_α one has that ψ_α is also a spin $1/2$ object (so that their product is a scalar), which can be identified with the usual chiral fermion. Given that, we can deduce the mass dimensionality of θ: $[\theta] = [\Phi]-[\psi] = [\phi]-[\psi] = -1/2$. Lastly, the coefficient of the θ^2 term must also be a (complex) scalar. (Note that if it was a vector field, which could also allow for the total spin of this term to be zero, it would have carried an additional space-time index, unlike Φ.) Its mass dimension is also uniquely determined $[F] = [\Phi] - 2[\theta] = 2$.

The coefficient $F(x)$ cannot be identified with any known physical $d.o.f.$ (There are no fundamental scalars whose mass dimension equals two!) Furthermore, its presence implies four bosonic $vs.$ only two fermionic $d.o.f.$, seemingly violating the supersymmetric principle found in the previous chapter. In order to understand the role of F and the resolution of this issue, let us first find the (supersymmetry) transformation law.

Consider the transformation $S(\Theta, \bar\Theta)$ acting on the superfield $\Phi(x, \theta)$ of eq. (4.6) (where Θ is also a Grassmann variable). Listing all objects at our disposal with dimensionality greater than $[\phi] - [\Theta] = 3/2$ but less than $[F] - [\Theta] = 5/2$ we have a spin 1/2 mass-dimension 3/2 object ψ; spin zero mass-dimension 2 objects F and $\partial_\mu \phi$; and a spin 1/2 mass-dimension 5/2 object $\partial_\mu \psi$. Using the Pauli matrices to contract space-time and spinorial indices, and matching spin and dimensions of the different objects O_I: $[\delta O_I] = [\Theta] + [O_J]$, one immediately arrives at the transformation law (up to the bracketed coefficients)

$$\delta\phi = \Theta^\alpha \psi_\alpha, \tag{4.7}$$

$$\delta\psi_\alpha = (2)\Theta_\alpha F + (2i)\sigma^\mu_{\dot\alpha\alpha}\bar\Theta^{\dot\alpha}\partial_\mu\phi, \tag{4.8}$$

$$\delta F = (-i)(\partial_\mu\psi^\alpha)\sigma^\mu_{\dot\alpha\alpha}\bar\Theta^{\dot\alpha}. \tag{4.9}$$

The "component fields" transform into each other, which is possible since the superspace coordinate carries mass dimensions and spin. Hence, the superfield Φ transforms onto itself.

The (linearized) transformation law can be written in a more compact fashion:

$$S(\Theta, \bar\Theta)\Phi(x, \theta) = \left\{\Theta\frac{\partial}{\partial\theta} + \bar\Theta\frac{\partial}{\partial\bar\theta} + 2i\theta\sigma^\mu\bar\Theta\partial_\mu\right\}\Phi(x, \theta), \tag{4.10}$$

where the second term on the right-hand side is zero and the identity (4.5) was used to derive (4.9) (from the last term on the right-hand side). Rewriting the differential transformation in a standard form $S(\Theta, \bar\Theta)\Phi(x, \theta) = \left\{\Theta Q + \bar Q \bar\Theta\right\}\Phi(x, \theta)$, where Q is the supercharge, $i.e.$, the generator of the supersymmetric algebra, on can identify the generators

$$Q_\alpha = \frac{\partial}{\partial\theta_\alpha}; \quad \bar Q_{\dot\alpha} = -\frac{\partial}{\partial\bar\theta_{\dot\alpha}} + 2i\theta^\alpha\sigma^\mu_{\dot\alpha\alpha}\partial_\mu, \tag{4.11}$$

and $P_\mu = i\partial/\partial x^\mu = i\partial_\mu$.

The generators are fermions rather than bosons! Their anticommutation relations can be found by performing successive transformations on Φ. After carefully re-arranging the result, it reads

$$\left\{Q_\alpha, \bar Q_{\dot\beta}\right\} = 2\sigma^\mu_{\alpha\dot\beta}P_\mu, \tag{4.12}$$

$$\{Q_\alpha, Q_\beta\} = \left\{\bar Q_{\dot\alpha}, \bar Q_{\dot\beta}\right\} = 0. \tag{4.13}$$

Relation (4.12) implies that in order to close the algebra, one needs to include space-time. (This hints towards a relation to gravity as is indeed one finds in the local case.) Thus, the algebra must also include the usual commutation relation

$$[P_\mu, P_\nu] = 0 \tag{4.14}$$

as well as the relations

$$[Q_\alpha, P_\nu] = [\bar Q_{\dot\alpha}, P_\nu] = 0. \tag{4.15}$$

The last relation (4.15) is only the statement that the supersymmetry is a global symmetry. (More generally, supersymmetry provides a unique consistent extension of the Poincare group.) Such an algebra which includes both commutation and anti-commutation relations among its generators is referred to as a graded Lie algebra.

The algebra described here has one supersymmetric charge, Q. Hence, this is a $N = 1$ (global) supersymmetry. In general, theories with more supersymmetries can be constructed. However, already $N = 2$ theories do not contain chiral matter, and $N = 4$ theories contain only gauge fields. Hence, extended supersymmetric theories, with certain exceptions in the case of $N = 2$, cannot describe nature at electroweak energies and $N = 1$ is unique in this sense. Discussion of the $N = 2$ case is reserved for the last part of these notes, and $N > 2$ theories will not be considered.

Though it is not our intention to discuss the algebra in great detail, some general lessons can be drawn by observation. The commutation between the charge and the Hamiltonian $[Q, H] = [Q, P_0] = 0$ implies that supersymmetric transformations which transform bosons to fermions and vice versa, do not change the "energy levels", $i.e.$, there is a boson-fermion mass degeneracy in these theories. The Hamiltonian can be written as $H = P_0 = (1/2)\{Q, \bar{Q}\} = QQ^\dagger \geq 0$ and hence is semi-positive definite. (That is, the potential V must vanish, $\langle V \rangle = 0$.) If (global) supersymmetry is conserved then the vacuum is invariant under the transformation $Q|0\rangle = 0$ and hence the vacuum energy $\langle 0|H|0 \rangle = 0$ is zero, regardless whether, for example, the gauge symmetry is conserved or not. If (global) supersymmetry is broken spontaneously then the vacuum energy $\sim \||Q|0\rangle\|^2 > 0$ is positive. The semi-positive definite vacuum energy is an order parameter of (global) supersymmetry breaking. The Goldstone particle of supersymmetry is given by the operation of its (super)current $\sim Q_\alpha$ on the vacuum, and hence it is a fermion ψ_α, the Goldstino.

The superfield $\Phi(x_\mu, \theta_\alpha)$ is called a chiral superfield since it contains a chiral fermion. It constitutes a representation of the algebra as is evident from the closed relations that we were able to deduce, mostly by simple arguments. (The arbitrary coefficients in (4.7)–(4.9) were fixed by requiring consistency of the algebra.) It is called the left- (or L-)representation. The more general chiral representation, which contains both chiral (left-handed) and anti-chiral (right-handed) fermions, is related to it by $\Phi(y_\mu, \theta_\alpha, \bar{\theta}_{\dot\alpha}) = \Phi_L(y_\mu + i\theta_\alpha \sigma_\mu^{\alpha\dot\alpha} \bar{\theta}_{\dot\alpha}, \theta_\alpha) [\equiv \Phi_L(x_\mu, \theta_\alpha)]$, where the shift in y_μ is to be treated as a translation operation on the superfield, and the object that was previously denoted by Φ is now denoted by Φ_L. There is also a right (or R-)representation given by $\Phi(y_\mu, \theta_\alpha, \bar{\theta}_{\dot\alpha}) = \Phi_R(y_\mu - i\theta_\alpha \sigma_\mu^{\alpha\dot\alpha} \bar{\theta}_{\dot\alpha}, \bar{\theta}_{\dot\alpha}) [\equiv \Phi_R(x_\mu^*, \bar{\theta}_{\dot\alpha})]$. In these notes we will mostly use the L-representation.

It is instructive to verify that the L-representation is closed. For example, we saw that $S(\Theta, \bar{\Theta})\Phi_L(x_\mu, \theta) = \Phi_L'(x_\mu, \theta)$. Similarly, it is straightforward to show that $\Phi^2 = \phi^2 + 2\theta\phi\psi + \theta^2(2\phi F - (1/2)\psi^2)$ is also a left-handed field, where $\theta^2 = \theta^\alpha \theta_\alpha$ ($\bar{\theta}^2 = \bar{\theta}_{\dot{\alpha}} \bar{\theta}^{\dot{\alpha}}$). Also, multiplying $\Phi^2 \Phi$ one has $\Phi^3 = \cdots + \theta^2 (\phi^2 F - (1/2)\phi\psi^2)$, a result which will be useful below.

Before returning to discuss the dimension two scalar field that seems to have appeared in the spectrum, the vector representation (which also contains such a field) will be given.

4.3 The Vector Superfield

A vector field $V_\mu(x_\mu)$ is to be embedded in a real object, a vector superfield $V(x_\mu, \theta_\alpha, \bar{\theta}_{\dot{\alpha}}) = V^\dagger(x_\mu, \theta_\alpha, \bar{\theta}_{\dot{\alpha}})$. Hence, aside from vector bosons, it can contain only a real scalar configuration and Majorana fermions. If it to contain spin zero and/or spin 1/2 fields, it cannot carry itself a space-time index. Also, it must depend on real combinations of θ and $\bar{\theta}$. In particular, its expansion contains terms up to a $\theta^2 \bar{\theta}^2$ term. One then arrives at the following form (coefficients are again fixed by the consistency of the algebra)

$$V(x_\mu, \theta_\alpha, \bar{\theta}_{\dot{\alpha}}) = -\theta_\alpha \sigma_\mu^{\alpha\dot{\alpha}} \bar{\theta}_{\dot{\alpha}} V^\mu(x_\mu)$$

$$+i\theta^2 \bar{\theta}^{\dot{\alpha}} \bar{\lambda}_{\dot{\alpha}}(x_\mu) - i\bar{\theta}^2 \theta^\alpha \lambda_\alpha(x_\mu) + \frac{1}{2}\theta^2 \bar{\theta}^2 D(x_\mu). \quad (4.16)$$

The vector superfield V is a dimensionless spin zero object! Its component fields include, aside from the vector boson, also a Majorana fermion λ (the gaugino) and a real scalar field D, again with mass dimension $[D] = 2$. It again contains four bosonic $vs.$ only two fermionic $d.o.f.$, a problem we already encountered in the case of a chiral superfiled, and again we postpone its resolution.

In fact, the expansion (4.16) is not the most general possible expansion. The following terms: C, $i\theta\chi - i\bar{\theta}\bar{\chi}$, and $i\theta^2(A + iB) - i\bar{\theta}^2(A - iB)$, where A, B, C are real scalar fields and χ is again a Majorana fermion, are dimensionless, spin zero, real terms that could appear in V (in addition to other derivative terms). We implicitly eliminated the additional $d.o.f.$ by fixing the gauge. While in ordinary gauge theory the gauge parameter is a real scalar field, in supersymmetry it has to be a chiral superfiled, hence it contains four bosonic and two fermionic $d.o.f.$ Subtracting one bosonic $d.o.f.$ to be identified with the usual gauge parameter, three bosonic and two fermionic $d.o.f.$ are available for fixing the "supersymmetric gauge". These can then be used to eliminate A, B, C and χ from the spectrum. This choice corresponds to the convenient Wess-Zumino gauge, which is used in these notes.

Similar considerations to the ones used above in order to write the chiral field transformation laws (4.7)–(4.8) can be used to write the transformation law $S(\Theta, \bar{\Theta})V(x_\mu, \theta, \bar{\theta})$ of the component field of the vector superfiled,

$$\delta V_\mu = (i) \left[\Theta \sigma_\mu \bar{\lambda} + \bar{\Theta} \sigma_\mu \lambda \right], \quad (4.17)$$

$$\delta\lambda = \bar{\Theta}D + \Theta\sigma^{\mu\nu}\left[\partial_\mu V_\nu - \partial_\nu V_\mu\right], \tag{4.18}$$

$$\delta D = \sigma^\mu \partial_\mu \left[-\Theta\bar{\lambda} + \bar{\Theta}\lambda\right], \tag{4.19}$$

where spinorial indices are suppressed and $\sigma^{\mu\nu} = (i/2)[\sigma^\mu, \sigma^\nu]$, as usual.

An important object that transforms in the vectorial representation is $\Phi_L \Phi_L^\dagger$. In order to convince oneself, simply note that it is a real field. In more detail, note that $\Phi_L^\dagger$ must depend on $\bar{\theta}$ only, and is therefore in the right-handed representation $[\{\Phi_L(y_\mu, \theta, \bar{\theta})\}^\dagger = \{\Phi_L(x_\mu, \theta)\}^\dagger = \Phi_L^\dagger(x_\mu^*, \bar{\theta})]$. Hence, $\Phi_L \Phi_L^\dagger$ expansion contains terms up to $\theta^2 \bar{\theta}^2$. Of particular interest to our discussion below is the coefficient of this last term in the $\Phi_L \Phi_L^\dagger$ expansion, the D-term, $\Phi_L \Phi_L^\dagger = \cdots + \theta^2 \bar{\theta}^2 D$ where

$$\int d^2\theta d^2\bar{\theta}\,\Phi_L \Phi_L^\dagger = D = FF^* - \phi\partial^\mu\partial_\mu\phi^* + (i/2)\psi\sigma_\mu\partial^\mu\bar{\psi}. \tag{4.20}$$

In turn, it is also possible to construct a useful chiral object W_α from vectorial fields (which carries spin $1/2$). For a $U(1)$ theory one has $W_\alpha = (\partial_{\bar{\theta}})^2(\partial_\theta + 2i\sigma_\mu\bar{\theta}\partial^\mu)V \sim \lambda_\alpha + \theta_\alpha(F_{\mu\nu} + D) + \theta^2\partial_\mu\lambda_\alpha$, and $F_{\mu\nu}$ is as usual the stress tensor. It is straightforward to show that W_α is a chiral field: From the anti-commutation one has $\partial_{\bar{\theta}}W_\alpha = 0$. It follows that W_α is independent of $\bar{\theta}$ and hence is a left-handed chiral superfield.

4.4 From the Auxiliary F and D Fields to the Lagrangian

After discussing chiral and vector superfields, understanding their expansion in terms of component fields and learning some simple manipulations of these super-objects, we are in position to do the final count of the physical $d.o.f.$ and resolve the tantalizing access of bosonic $d.o.f.$ which are manifested in the form of dimension-two scalar fields. This requires us to derive the usual interaction and kinetic terms for the component fields. We begin with a non-gauge theory.

It is not surprising that the two issues are closely related. Observe that the F-term of the bilinear Φ^2, given by $\int d^2\theta\Phi^2$, includes terms which resemble mass terms, while $\int d^2\theta\Phi^3$ includes Yukawa-like terms. Also, δF given in (4.9) is a total derivative, for any chiral superfield. That is, the mysterious F-field contains interaction and mass terms on the one hand, and, on the other hand, $\delta \int d^4xF = \int d^4x(\text{total} - \text{derivative}) = 0$ so that $\int d^4xF$ is invariant under supersymmetric transformations and hence, is a good candidate to describe the potential. Indeed, if it is a non-propagating (auxiliary) $d.o.f.$ which encodes the potential then there are no physical spin-two bosonic $d.o.f.$

Clearly, one needs to identify the kinetic terms in order to verify this conjecture. A reasonable guess would be that they are given by the self-conjugate combination $\Phi\Phi^\dagger$. Indeed, the corresponding D-term eq. (4.20) contains kinetic terms for ϕ and ψ, but not for F. The D-field itself also transforms (for any vector superfield) as a total derivative, so again it is not a coincidence that it is a term which could be consistently used to write the appropriate kinetic terms in the Lagrangian. The F-fields are therefore auxiliary non-propagating fields that could be eliminated from the theory. That is, they do not correspond to any physical $d.o.f.$ (But what about the D-fields? Are they also auxiliary fields? The kinetic terms of the gauge-fields are given by the F-term of W^2, which can be shown to not contain a kinetic term for the D-field: The D-term is again an auxiliary field which does not represent a physical $d.o.f.$)

More specifically, consider a theory described by a Lagrangian

$$\mathcal{L} = \int d^2\theta \left[\mu\Phi^2 + y\Phi^3\right] + h.c. + \int d^2\theta d^2\bar{\theta}\Phi\Phi^\dagger, \tag{4.21}$$

We can then describe the component (ordinary) field theory by performing the integrations $\int d^2\theta$ and $\int d^2\theta d^2\bar{\theta}$ over the super-space coordinates and extracting the corresponding F and D terms, respectively:

$$F_{\Phi^2} + F_{\Phi^2}^* : 2\mu\phi F - \frac{1}{2}\mu\psi\psi + h.c.; \tag{4.22}$$

$$F_{\Phi^3} + F_{\Phi^3}^* : 3y\phi^2 F - \frac{3}{2}y\phi\psi\psi + h.c.; \tag{4.23}$$

$$D \qquad : |\partial_\mu\phi|^2 - \frac{i}{2}\bar{\psi}\partial\!\!\!/\psi - FF^*; \tag{4.24}$$

The Lagrangian $\mathcal{L} = T - V$ is given by their sum.

Indeed, no kinetic terms appear for F. Hence, one can eliminate the auxiliary $d.o.f.$ by solving $\partial V/\partial F = 0$ and substituting the solution,

$$\frac{\partial V}{\partial F} = 0 \Rightarrow F^* = -(2\mu\phi + 3y\phi^2), \tag{4.25}$$

back into the potential V. One finds for the scalar potential (after reorganizing the conjugate terms)

$$V(\phi) = FF^* = 4|\mu|^2\phi\phi^* + 6(y\mu^*\phi^*\phi^2 + h.c.) + 9|y|^2|\phi\phi^*|^2. \tag{4.26}$$

The fermion mass, kinetic and (Yukawa) interaction terms are already given explicitly in (4.22)–(4.24).

It is now instructive to compare our result (4.26) to our conjectured "finite" theory from the previous chapter. In order to do so it is useful to use standard normalizations, $i.e.$, $\mu \to \mu/2$; $y \to y/3$; and most importantly $\psi \to \psi/\sqrt{2}$. Not surprisingly, our Lagrangian includes all of (and only) the $d.o.f.$ previously conjectured and it exhibits the anticipated structure of interactions. In particular trilinear and quartic couplings are not arbitrary and are given by the masses and Yukawa couplings, as required. Quartic couplings are (semi-)positive definite.

4.5 The Superpotential

Let us now define a function

$$W = \mu\Phi^2 + y\Phi^3, \tag{4.27}$$

then $F^* = -\partial W/\partial\Phi$ where the substitution $\Phi \to \phi$ is understood at the last step. The scalar potential is then simply

$$V = |\partial W/\partial\Phi|^2.$$

Also note that the potential terms involving the fermions are simply given by

$$-(1/2)\sum_{IJ}(\partial^2 W/\partial\Phi_I\partial\Phi_J)\psi_I\psi_J + h.c.$$

(These general "recipes" will not be justified here.) The function W is the superpotential which describes our theory. In general, the superpotential could be any analytic (holomorphic) function. In particular, it can depend on any number of chiral superfields Φ_I, but not on their complex conjugates $\Phi_I^\dagger$ (the holomorphicity property).

For example, the most general renormalizable superpotential describing a theory with one singlet field is $W = c + l\Phi + \mu\Phi^2 + y\Phi^3$. That is, W has mass dimension $[W] = 3$ and any terms which involve higher powers of Φ are necessarily non-renormalizable terms. Indeed, $W = \kappa\Phi^4 \Rightarrow V = |\partial W/\partial\Phi|^2 = 8\kappa^2\phi^6$, *i.e.*, $[\kappa] = -1$ is a non-renormailzable coupling. The holomorphicity property of W is at the core on the non-renormalization theorems that protect its parameters from divergences (and of which some aspects were shown in the previous lecture but from a different point of view).

4.6 The Kähler Potential

The non-gauge theory is characterized by its superpotential and also by its kinetic terms. Above, it was shown that the kinetic terms are given in the one singlet model simply by $\Phi\Phi^\dagger$. More generally, however, one could write a functional form

$$\Phi\Phi^\dagger \to K(\Phi, \Phi^\dagger) = K_J^I\Phi_I\Phi^J + \left(H^{IJ}\Phi_I\Phi_J + h.c.\right)$$
$$+\text{Non-renormalizable terms} \tag{4.28}$$

(with $\Phi^J = \Phi_J^\dagger$) which is a dimension $[K] = 2$ real function: The Kähler potential. The canonical kinetic terms $K_J^I\partial_\mu\phi_I\partial^\mu\phi^J = \sum_I |\partial_\mu\phi_I|^2$ are given by the "minimal" Kähler potential, which is a renormalizable function with $K^I_J = \delta^I_J$, as we implicitly assumed above. Note that holomorphic terms (*i.e.*, terms which do not depend on $\Phi^\dagger$'s) such as $H^{IJ}\Phi_I\Phi_J$ do not alter the normalization of the kinetic terms since $\int d^2\theta d^2\bar\theta \left\{ H^{IJ}\Phi_I\Phi_J + h.c. \right\} = 0$. (These terms, however, could play an interesting role in the case of local supersymmetry, *i.e.*, supergravity.)

The Kähler potential which determines the normalization of the kinetic terms cannot share the holomorphicity of the superpotential. Hence, W and K are very different functions which together define the (non-gauge) theory. In particular, the wave function renormalization of the fields (and therefore of the superpotential couplings) is determined by the Kähler potential which undergoes renormalization (or scaling).

4.7 The Case of a Gauge Theory

The more relevant case of a gauge theory is obviously more technically evolved. Here, we will only outline the derivation of the component Lagrangian of an interacting gauge theory, stressing those elements which will play a role in the following chapters. The pure gauge theory is described by

$$\int d^2\theta \frac{1}{2g^2} W^\alpha W_\alpha, \qquad (4.29)$$

leading to the usual gauge kinetic terms, as well as kinetic terms for the gaugino λ, given by (in the normalization of eq. (4.29))

$$F_{W^2} : -\frac{1}{4}|F^{\mu\nu}|^2 + \frac{1}{2}D^2 - \frac{i}{2}\bar\lambda \slashed{D}\lambda, \qquad (4.30)$$

with $\slashed{D}$ as usual the derivative operator contracted with the Pauli matrices. There are kinetic term for the physical gauge bosons and fermions, but not for the D fields. The D fields are indeed auxiliary fields.

More generally, $(1/2g^2)W^\alpha W_\alpha \to f_{\alpha\beta}W^\alpha W^\beta$ and $f_{\alpha\beta}$ is a holomorphic (analytic) function which determines the normalization of the gauge kinetic terms: $f_{\alpha\beta}$ is the gauge-kinetic function. It could be a constant pre-factor or a function of a dynamical (super)field $f_{\alpha\beta}(S)$. S is then the dilaton superfield. In either case (once put in its canonical form) it is essentially equivalent to the gauge coupling, $f_{\alpha\beta} = \delta_{\alpha\beta}/2g^2$, only that in the latter case the gauge couplings is determined by the *vev* of the dilaton S.

As before, the usual kinetic terms are described by the D-terms, however, they are now written as $\int d^2\theta d^2\bar\theta \Phi e^{2gQV}\Phi^\dagger$ where for simplicity the gauge coupling will be set $g = 1$ (see exercise); $Q = 1$ is the charge, and $\Phi e^{2V}\Phi^\dagger = \Phi\Phi^\dagger + \Phi 2V\Phi^\dagger + \cdots$ contains the gauge-covariant kinetic terms ($\partial_\mu \to D_\mu \equiv \partial_\mu + iV_\mu$). (More generally, $\Phi e^{2gQV}\Phi^\dagger \to K(\Phi_I, e^{2gQ_JV}\Phi^J)$.) That is,

$$D : |D_\mu\phi|^2 - \frac{i}{2}\bar{\psi}\not{D}\psi + \phi^* D\phi + i\phi^*\lambda\psi - i\bar{\lambda}\bar{\psi}\phi - FF^*. \qquad (4.31)$$

Note that when recovering the explicit dependence on the gauge coupling, the strength of the gauge Yukawa interaction $\phi^*\lambda\psi - \bar{\lambda}\bar{\psi}\phi$ is g. (In the standard normalization $\psi \to \psi/\sqrt{2}$ it is $\sqrt{2}g$.) Note that the Lagrangian of the toy theory described in Sect. 3.1 is derived here from the gauge Lagrangian eq. (4.29), as was already suggested following eq. (3.5).

Consider now a theory

$$\int d^2\theta \left[(\mu\Phi^+\Phi^- + h.c.) + \frac{1}{2}W^\alpha W_\alpha\right] + \int d^2\theta d^2\bar{\theta} \left\{\Phi^+ e^{2V}\Phi^{+\dagger} + \Phi^- e^{-2V}\Phi^{-\dagger}\right\},$$
$$(4.32)$$

where the matter fields carry $+1$ and -1 charges. One has

$$\frac{\partial V}{\partial F^+} = 0 \Rightarrow F^{+*} = -\mu\phi^- \qquad (4.33)$$

$$\frac{\partial V}{\partial F^-} = 0 \Rightarrow F^{-*} = -\mu\phi^+ \qquad (4.34)$$

$$\frac{\partial V}{\partial D} = 0 \Rightarrow -D = \phi^{+*}\phi^+ - \phi^{-*}\phi^-, \qquad (4.35)$$

where in deriving (4.35) we used eqs. (4.30) and (4.31). The scalar potential is now readily derived,

$$V(\phi) = FF^* + \frac{1}{2}D^2 = |\mu|^2\phi^+\phi^{+*} + |\mu|^2\phi^-\phi^{-*} + \frac{1}{2}|\phi^+\phi^{+*} - \phi^-\phi^{-*}|^2. \quad (4.36)$$

The quartic potential is now dictated by the gauge coupling ($-D = g(\phi^{+*}\phi^+ - \phi^{-*}\phi^-)$) once reinstating the gauge coupling) and again is semi-positive definite. Note that the quartic D-potential (the last term in (4.36)) has a flat direction $\langle\phi^+\rangle = \langle\phi^-\rangle$, possibly destabilizing the theory: This is a common phenomenon is supersymmetric field theories.

The derivation of the interactions in the case of a gauge theory essentially completes our sketch of the formal derivation of the ingredients that were required in the previous section in order to guarantee that the theory is at most logarithmically divergent. Specifically, quartic and gaugino couplings were found to be proportional to the gauge coupling (or its square) and with the appropriate coefficients. One is led to conclude that supersymmetry is indeed a natural habitat of weakly coupled theories. (Supersymmetry also provides powerful tools for the study of strongly coupled theories [47], an issue that will not be addressed in these notes.) The next step is then the supersymmetrization of the Standard Model. Beforehand, however, we comment on gauging supersymmetry itself (though for the most part these notes will be concerned with global supersymmetry).

4.8 Gauging Supersymmetry

The same building blocks are used in the case of local supersymmetry, which is referred to as supergravity, only that the gravity multiplet (which includes the graviton and the spin 3/2 gravitino) is now a gauge multiplet. The gravitino absorbs the Goldstino once supersymmetry is spontaneously broken and acquires a mass

$$m_{3/2} = \frac{\langle W \rangle}{M_P^2} e^{\frac{\langle K \rangle}{2 M_P^2}}. \tag{4.37}$$

The reduced Planck mass $M_P = M_{\text{Planck}}/\sqrt{8\pi} \simeq 2.4 \times 10^{18}$ GeV is the inverse of the supergravity expansion parameter (rather than the Planck mass itself) and it provides the fundamental mass scale in the theory. Hence, local supersymmetry corresponds to gauging gravity (and hence, supergravity). The appearance of momentum in the anticommutation relations of global supersymmetry already hinted in this direction. Gravity is now included automatically, opening the door for gauge-gravity unification. However, supergravity is a non-renormalizable theory and is still expected to provide only a "low-energy" limit of the theory of quantum gravity (for example, of a (super)string theory).

The gravitino mass could be given by an expectation value of any superpotential term, particularly a term that does not describe interactions of known light fields. Because supergravity is a theory of gravity, all fields in all sectors of the theory "feel" the massive gravitino (since (super)gravity is modified). For example, one often envisions a scenario in which supersymmetry is broken spontaneously in a hidden sector – hidden in the sense that it interacts only gravitationally with the SM sector – and the SM sector appears as globally supersymmetric but with explicit supersymmetry breaking terms that are functions of the gravitino mass. (See Sect. 10.2.)

The F-terms are still good order parameters for supersymmetry breaking though they take a more complicated form (since they are now given by a covariant derivative of the superpotential):

$$-F_\Phi^* = \frac{\partial W}{\partial \Phi} + \frac{\partial K}{\partial \Phi} \frac{W}{M_P^2}, \tag{4.38}$$

where in the canonical normalization one has $\partial K/\partial \Phi = \phi^*$. Indeed, $|F| > 0$ if supersymmetry is broken. The scalar potential (or vacuum energy) now reads

$$V = e^{K/M_P^2} \left\{ F_I K_J^I F^J - \frac{3}{M_P^2} |W|^2 \right\}, \tag{4.39}$$

and is not a good order parameter as it could take either sign, or preferably vanish, even if $|F| > 0$. This is in fact a blessing in disguise as it allows one to cancel the cosmological constant (that was fixed in the global case to $\langle V \rangle = \langle |F|^2 \rangle$) by tuning

$$\langle W \rangle = \frac{1}{\sqrt{3}} \langle F \rangle M_P. \tag{4.40}$$

One can define the scale of spontaneous supersymmetry breaking

$$M_{SUSY}^2 = \langle F \rangle \exp[\langle K \rangle / 2M_P^2]. \tag{4.41}$$

The cancellation of the cosmological constant then gives

$$M_{SUSY}^2 = \sqrt{3} m_{3/2} M_P \tag{4.42}$$

as a geometric mean of the gravitino and supergravity scales. Note that regardless of the size of M_{SUSY}, supersymmetry breaking is communicated to the whole theory (to the full superpotential) at the supergravity scale M_P, as is evident from the form of the potential (4.39). (The symbol M_{SUSY} itself is also used to denote the mass scale of the superpartners of the SM particles, as we discuss in the following chapter. In that case M_{SUSY} is the scale of explicit breaking in the SM sector. The two meaning should not be confused.)

There could be other scales $M < M_P$ at which other mechanisms mediate supersymmetry breaking from some special (but by definition not truly hidden) sector of the theory to the (supersymmetrized) SM fields. In this case there is multiple mediation and one has to examine whether supergravity or a different "low-energy" mechanism dominates. (We will return to this point when discussing models of supersymmetry breaking in Chaps. 10 and 13.)

Of course, the supergravity scalar potential has also the usual D-term contribution $V_D = (1/4 f_{\alpha\beta}) D_\alpha D_\beta$, and the D-terms are also order parameters of supersymmetry breaking. (However, it seems more natural to fine-tune the supepotential to cancel its first derivative (the F-term) in order to eliminate the cosmological constant, rather than fine-tune it to cancel the square of the D-terms.)

We will briefly return to supergravity when discussing in Chap. 10 the origins of the explicit supersymmetry breaking in the SM sector, but otherwise the basic tools of global supersymmetry will suffice for our purposes.

Exercises

4.1 Since Φ_L by itself constitutes a representation of the algebra, its component fields cannot transform for example to a vectorial object V_μ. Show this explicitly from the consideration of dimensions, space-time and spinorial indices.

4.2 Use the definition of Φ_R to show that it is independent of θ_α.

4.3 Confirm the expansion of Φ_L^2 and Φ_L^3.

4.4 Examine the expansion of the vector superfield (4.16), show that $V = V^\dagger$, and confirm all the assertions regarding spin and mass dimension of the component fields.

4.5 Derive the transformation law (4.17)–(4.19) (aside from coefficients) from dimensional and spin arguments.

4.6 Use the relations given in the text to derive the expression for $\Phi(y_\mu, \theta, \bar{\theta}) = \Phi_L(y_\mu + i\theta\sigma_\mu\bar{\theta}, \theta) = [1 - \delta y_\mu P^\mu + (1/2)(\delta y_\mu P^\mu)^2]\Phi_L(y_\mu, \theta)$ with $\delta y_\mu = i\theta\sigma_\mu\bar{\theta}$. Derive the expression for $\Phi_L^\dagger$. Integrate $\int d^2\theta d^2\bar{\theta}\Phi_L\Phi_L^\dagger$ to find the $\Phi_L\Phi_L^\dagger$ D-term.

4.7 Derive the complete expression for the vector field in its spinorial-chiral representation W_α and confirm explicitly that it is a spin $1/2$ chiral superfield.

4.8 Verify that our derivation of the non-gauge potential agrees with the theory conjectured in the previous chapter.

4.9 Find the correctly normalized Yukawa couplings y, $W = (y/3)\Phi_I^3$, if $K_J^I = k_I\delta_J^I$. (The fields have to be rescaled so that the kinetic terms have their canonical form.) What about $W = y_{IJK}\Phi_I\Phi_J\Phi_K$? Verify your choice by examining the resulting form of the quartic potential.

4.10 Show that $\int d^2\theta d^2\bar{\theta}\left\{H^{IJ}\Phi_I\Phi_J + h.c.\right\} = 0$.

4.11 Recover the explicit dependence on the gauge coupling and on the charge in the expressions in Sect. 4.7. Show that the quartic coupling is proportional to $g^2/2$ and that self couplings are positive definite.

4.12 Introduce a Yukawa term $y\Phi^0\Phi^+\Phi^-$ to the model described by eq. (4.32) and work out the scalar potential and the component Yukawa interactions. What are the quartic couplings?

4.13 Show that the Lagrangians (4.21), (4.29) and (4.32) are total derivatives.

4.14 Show that all the supergravity expressions reduce to the global supersymmetry case in the limit $M_P \to \infty$ in which supergravity effects decouple. In this limit, for example, supersymmetry breaking is not communicated from the hidden to the SM sector. It may be communicated in this case from some other sector via means other than gravity.

5. Supersymmetrizing the Standard Model

The discussion in the previous sections suggests that the SM could still be a sensible theory at the ultraviolet and at the same time insensitive to the ultraviolet cut-off scale if supersymmetry is realized in the infrared regime (specifically, near the Fermi scale). Let us then consider this possibility and construct a (minimal) supersymmetric extension of the SM.

5.1 Preliminaries

Nearly all of the necessary ingredients are already present within the SM: scalar bosons, (chiral) fermions and gauge bosons. Each of the SM gauge-boson multiplets requires the presence of a real (Majorana) fermion with the same quantum numbers, its *gaugino* superpartner; each SM fermion requires the presence of a complex scalar boson with the same quantum numbers, its *sfermion* superpartner (which also inherits its chirality label); the Higgs boson requires the introduction of the *Higgsino*, its fermion superpartner. It is straightforward to convince oneself that the theory will not be anomaly free unless two Higgsino doublets (and hence, by supersymmetry, two Higgs doublets) with opposite hypercharge are introduced so that the trace over the Higgsino hypercharge vanishes, in analogy to the vanishing hypercharge trace of each SM generation. (Recall our construction in Chap. 3.) The minimal supersymmetric extension of the Standard Model (MSSM) as was outlined above is the extension with minimal new matter content, and it must contain a two Higgs doublet model (2HDM) (see eq. (3.3)) with $H_1 = ((H_1^0 + iA_1^0)/\sqrt{2}, H_1^-)^T$ and $H_2 = (H_2^+, (H_2^0 + iA_2^0)/\sqrt{2})^T$. Note that the MSSM contains also Majorana fermions. Discovery of supersymmetry will not only establish the existence of fundamental scalar fields but also of Majorana fermions!

We list, following our previous notation, the MSSM field content in Table 5.1. For completeness we also include the gravity multiplet with the spin 2 graviton G and its spin 3/2 gravitino superpartner $\widetilde{G}$, as well as a generic supersymmetry breaking scalar superfield X (with an auxiliary component $\langle F_X \rangle \neq 0$) whose fermion partner $\widetilde{X}$ is the Goldstone particle of supersymmetry breaking, the Goldstino. The Goldstino is absorbed by the gravitino. (X

parameterizes, for example, the hidden sector mentioned in Sect. 4.8.) It is customary to name a sfermion with an "s" suffix attached to the name of the corresponding fermion, e.g. *top*-quark → *stop*-squark and τ-lepton → *stau*-slepton. The gaugino partner of a gauge boson V is typically named V-ino, for example, W-boson → wino. The naming scheme for the mass eigenstates will be discussed after the diagonalization of the mass matrices. All superpartners of the SM particles are denoted as s-particles, or simply *sparticles*.

Given the above matter content and the constrained relations between the couplings that follow from supersymmetry, the MSSM is guaranteed not to contain quadratic divergences. In order to achieve that and have a sensible formalism to treat a fundamental scalar we were forced to introduce a complete scalar replica of the SM matter which, however, supersymmetry enables us to understand in terms of a boson-fermion symmetry. By doing so, one gives up in a sense spin as a good quantum number. For example, the Higgs doublet H_1 and the lepton doublets L_a are indistinguishable at this level: In the SM the former is a scalar field while the latter are fermions. The anomaly cancellation considerations explained above do not allow us to identify H_1 with L_a, *i.e.*, the Higgs boson with a slepton (or equivalently, a lepton with a Higgsino). The possibility of lepton-Higgs mixing arises naturally if no discrete symmetries are imposed to preserve lepton number, a subject that we will return to below. Spin and the SM spectrum correspond in this framework to the low-energy limit. That is, supersymmetry must be broken at a scale below the typical cut-off scale that regulates the quadratic divergences $\Lambda \lesssim 4\pi M_W$. If the stop-squark, for example, is heavier, then divergences must be again fine-tuned away – undermining our original motivation. (As we shall see, consistency with experiment does not allow the stop to be much lighter.)

As explained in the previous section, the boson-fermion symmetry and the duplication of the spectrum follow naturally once the building blocks used to describe nature (at electroweak energies and above) are the (chiral and vector) superfields. However, from the low-energy point of view it is the *component field* interactions which are relevant. Let us then elaborate on the structure of the *component field* interactions, which is dictated by the supersymmetry invariant interactions of the parent superfields. This will be done in the next sections.

Before doing so, it is useful to fix the normalization of the Higgs fields, as their expectation values appear in essentially all mass matrices of the MSSM fields. It is customary to define in 2HDM in general, and in the MSSM in particular,

$$\sqrt{\langle H_1^0 \rangle^2 + \langle H_2^0 \rangle^2} = \tfrac{\sqrt{2}}{g} M_W \simeq 174 \text{ GeV} \equiv \nu.$$

Note that a $1/\sqrt{2}$ factor is now absorbed in the definition of H_i (and ν) in comparison to the Higgs doublet definition in the SM. The *vev* ν is not a free parameter but is fixed by the Fermi scale (or equivalently by the W or Z mass). However, the ratio[1] of the two $vev's$

$$\tan \beta = \frac{\langle H_2^0 \rangle}{\langle H_1^0 \rangle} \equiv \frac{\nu_2}{\nu_1} \gtrsim 1 \qquad (5.1)$$

is a free parameter. Its positive sign corresponds a conventional phase choice that we adopt, and its lower bound stems from pertubativity of Yukawa couplings $\propto 1/\sin \beta$. (See Chap. 7.) Correspondingly, one has $\langle H_1^0 \rangle \equiv \nu_1 = \nu \cos \beta$ and $\langle H_2^0 \rangle \equiv \nu_2 = \nu \sin \beta$.

The supersymmatrization of the standard model is also discussed in many of the reviews and notes listed in the previous chapters (particularly, Refs. [40, 37, 45]), as well as in Refs. [48], [49], and [50]. Recent summaries by the relevant Tevatron [51] and by other [52] working groups are also useful, but address particular models. Historical perspective on the supersymmetric SM was given recently by Fayet [53].

5.2 Yukawa, Gauge-Yukawa, and Quartic Interactions

Identifying the gauge kinetic function $f_{\alpha\beta} = \delta_{\alpha\beta}/2g_a^2$ for a gauge group a, our exercise in Sect. 4.7 showed that the fermion f coupling to a vector boson V_a with a coupling g_a implies, by supersymmetry, the following interaction terms:

$$g_a V_a \bar{f} T_a^i f \rightarrow \begin{cases} \sqrt{2} g_a \lambda_a \tilde{f}^* T_a^i f + h.c. \\[2ex] \frac{g_a^2}{2} \sum_I \left| \tilde{f}_I^* T_a^i \tilde{f}_I \right|^2 \end{cases},$$

in addition to usual gauge interactions of the sfermion $\tilde{f}$. Here, we explicitly denote the gauge group generators T_a^i which are taken in the appropriate representation (of the fermion f) and which are reduced to the fermion charge in the case of a $U(1)$ considered above. The gaugino is denoted by λ_a and, converting to standard conventions, it is defined to absorb a factor of i which appeared previously in the gaugino-fermion-sfermion vertex. Also, the fermion component of the chiral superfield absorbs a factor of $1/\sqrt{2}$ as explained in the previous section. The gaugino matter interaction is often referred to as gauge-Yukawa interaction. The quartic interaction is given by the square of the auxiliary D field and hence includes a summation over all

---

[1] The definition given here corresponds to a physical parameter only at tree level. The "physical" definition of the angle β (see also Chap. 8) is corrected at one-loop in a way which depends on the observable used to extract it.

Table 5.1. The Minimal Supersymmetric SM (MSSM) field content. Our notation is explained in Table 1.1. Each multiplet $(Q_c, Q_L)_{Q_{Y/2}}$ is listed according to its color, weak isospin and hypercharge assignments. Note that chirality is now associated, by supersymmetry, also with the scalar bosons. In particular, the superpartner of a fermion f_L^c is a sfermion $\tilde{f}_R^*$ and that of a fermion f_L is a sfermion $\tilde{f}_L$. The model contains two Higgs doublets. In the case of the matter (Higgs) fields, the same symbol will be used for the superfield as for its fermion (scalar) component.

Multiplet	Boson	Fermion
Gauge fields		
$(8, 1)_0$	g	$\tilde{g}$
$(1, 3)_0$	W	$\widetilde{W}$
$(1, 1)_0$	B	$\tilde{B}$
Matter fields		
$(3, 2)_{\frac{1}{6}}$	$\tilde{Q}_a$	Q_a
$(\bar{3}, 1)_{-\frac{2}{3}}$	$\tilde{U}_a$	U_a
$(\bar{3}, 1)_{\frac{1}{3}}$	$\tilde{D}_a$	D_a
$(1, 2)_{-\frac{1}{2}}$	$\tilde{L}_a$	L_a
$(1, 1)_1$	$\tilde{E}_a$	E_a
Symmetry breaking		
$(1, 2)_{-\frac{1}{2}}$	H_1	$\widetilde{H}_1$
$(1, 2)_{\frac{1}{2}}$	H_2	$\widetilde{H}_2$
Gravity and supersymmetry breaking		
$(1, 1)_0$	G	$\tilde{G}$
$(1, 1)_0$	X	$\tilde{X}$

the scalar fields which transform under the gauge group, weighted by their respective charge. This leads to "mixed" quartic interactions, for example, between two squarks and two Higgs bosons or two squarks and two sleptons.

The matter (Yukawa) interactions, and as we learned – the corresponding trilinear and quartic terms in the scalar potential, are described by the superpotential. Recall that the superpotential W is a holomorphic function so it cannot contain complex conjugate fields. Particularly, no terms involving H_1^* can be written. Indeed, we were already forced to introduce $H_2 \sim H_1^*$, for the purpose of anomaly cancellation. The holomorphicity property offers, however, an independent reasoning for introducing two Higgs doublets with opposite hypercharge: A Yukawa/mass term for the *up* quarks, which $\sim H_1^*$ in the SM, can now be written using H_2. Keeping in mind the generalization $V \sim y\phi_i\psi_j\psi_k \to W \sim y\Phi_I\Phi_J\Phi_K$, the MSSM Yukawa superpotential is readily written

$$W_{\text{Yukawa}} = y_{l_{ab}}\epsilon_{ij}H_1^i L_a^j E_b + y_{d_{ab}}\epsilon_{ij}H_1^i Q_a^j D_b - y_{u_{ab}}\epsilon_{ij}H_2^i Q_a^j U_b, \quad (5.2)$$

were $SU(2)$ indices are explicitly displayed (while $SU(3)$ indices are suppressed) and our phase convention is such that all "mass" terms and Yukawa couplings are positive.

It is instructive to explicitly write all the component interactions encoded in the superpotential (5.2). Decoding the $SU(2)$ structure and omitting flavor indices, one has

$$\begin{aligned} W_{\text{Yukawa}} = {} &y_l(H_1^0 E_L^- E - H_1^- N_L E) + y_d(H_1^0 D_L D - H_1^- U_L D) \\ &+y_u(H_2^0 U_L U - H_2^+ D_L U), \end{aligned} \quad (5.3)$$

with self explanatory definitions of the various superfield symbols. We note in passing that explicit decoding of the $SU(2)$ indices need to be carried out cautiously in certain cases, for example, when studying the vacuum structure (which is generally not invariant under $SU(2)$ rotations). However, it is acceptable here. Considering next, for example, the first term $y_l H_1^0 E_L^- E$, one can derive the component interaction $\sim (\partial^2 W/\partial\Phi_I\Phi_J)\psi_I\psi_J + h.c. + |\partial W/\partial\Phi_I|^2$ (where the substitution $\Phi = \phi + \theta\psi \to \phi$ in $\partial^n W$ is understood),

$$\mathcal{L}_{\text{Yukawa}} = y_l\left\{H_1^0 E_L^- E + \tilde{H}_1^0 \tilde{E}_L^- E + \tilde{H}_1^0 E_L^- \tilde{E}\right\} + h.c., \quad (5.4)$$

$$V_{\text{scalar(quartic)}} = y_l^2\left\{|\tilde{E}_L^- \tilde{E}|^2 + |H_1^0 \tilde{E}_L^-|^2 + |H_1^0 \tilde{E}|^2\right\}. \quad (5.5)$$

A single Yukawa interaction in the SM $H_1^0 E_L^- E$ leads, by supersymmetry, to two additional Yukawa terms and three quartic terms in the MSSM (or any other supersymmetric extension).

It is important to note that the holomorphicity of the superpotential which forbids terms $\sim H_1^*QU$, H_2^*QD, H_2^*LE, automatically implies that the MSSM contains a 2HDM of type II, that is a model in which each fermion flavor couples only to one of the Higgs doublets. This is a crucial requirement for suppressing dangerous tree-level contributions to FCNC from operators such as $QQUD$ which result from virtual Higgs exchange in a general 2HDM, but which do not appear in a type II model.

5.3 The Higgs Mixing Parameter

While no "supersymmetric" (*i.e.*, a holomorphic superpotential) mass involving the (SM) matter fields can be written, a Higgs mass term $\mu\epsilon_{ij}H_1^iH_2^j$ is gauge invariant and is allowed. The dimension $[\mu] = 1$ parameter mixes the two doublets and acts as a Dirac mass for the Higgsinos. The Dirac fermion and the Higgs bosons are then degenerate, as implied by supersymmetry, with mass $|\mu|$. The μ-parameter can carry an arbitrary phase. Hence, choosing $\mu > 0$, one has for the MSSM superpotential

$$
\begin{aligned}
W_{\text{MSSM}} &= W_{\text{Yukawa}} \pm \mu\epsilon_{ij}H_1^iH_2^j \\
&= y_{l_{ab}}\epsilon_{ij}H_1^iL_a^jE_b + y_{d_{ab}}\epsilon_{ij}H_1^iQ_a^jD_b \\
&\quad - y_{u_{ab}}\epsilon_{ij}H_2^jQ_a^jU_b \pm \mu\epsilon_{ij}H_1^iH_2^j.
\end{aligned}
\tag{5.6}
$$

Following the standard procedure it is straightforward to derive the new interaction and mass terms, which complement the usual Yukawa and quartic potential terms,

$$
\mathcal{L}_{\text{Dirac}} \quad = \mp\mu\widetilde{H}_1^-\widetilde{H}_2^+ \pm \mu\widetilde{H}_1^0\widetilde{H}_2^0 + h.c.,
\tag{5.7}
$$

$$
V_{\text{scalar(mass)}} = \mu^2\left(H_1H_1^\dagger + H_2H_2^\dagger\right),
\tag{5.8}
$$

$$
V_{\text{scalar(trilinear)}} = \pm\mu y_l H_2\widetilde{L}^\dagger\widetilde{E}^\dagger \pm \mu y_d H_2\widetilde{Q}^\dagger\widetilde{D}^\dagger \mp \mu y_u H_1\widetilde{Q}^\dagger\widetilde{U}^\dagger + h.c.,
\tag{5.9}
$$

where $SU(2)$ indices were also suppressed and the fermion mass terms were written in $SU(2)$ components, as it will be useful later when constructing the fermion mass matrices. Note that the trilinear terms arise from the cross terms in $|\partial W/\partial H_2|^2 = |\pm\mu H_1 - y_u QU|^2$ and $|\partial W/\partial H_1|^2 = |\pm\mu H_2 + y_l LE + y_d QD|^2$.

Could it be that $\mu \simeq 0$? Naively, one may expect that in this case the Higgsinos are massless and therefore Z-boson decays to a Higgsino pair $Z \to \widetilde{H}\widetilde{H}$ should have been observed at the Z resonance (from total or invisible width measurements, for example). In fact, once electroweak symmetry breaking effects are taken into account (see below) the two charged Higgsinos as well as one neutral Higgsino are degenerate in mass with the respective Goldstone bosons and are therefore massive with M_W and M_Z masses, respectively. Their mass follows from the gauge-Yukawa interaction terms $\sim g\langle H_i^0\rangle\widetilde{W}\widetilde{H}$ and $g\langle H_i^0\rangle\widetilde{Z}\widetilde{H}$. While the Z decays are kinematicly forbidden in this case, charged Higgsinos should have been produced in pairs at the WW production threshold at the Large Electron Positron (LEP) collider at CERN. The absence of anomalous events at the WW threshold allows one to exclude this possibility and to bound $|\mu|$ from below. (This observation will become clearer after defining the fermion mass eigenstates. For a discussion see Ref. [54].)

Obviously, $|\mu|$ cannot be too large or the Higgs doublets will decouple, re-introducing a conceptual version of the hierarchy problem. As we proceed we will see that μ must encode information on the ultraviolet in order for it to be a small parameter of the order of the Fermi scale. Hence, it may contain one of the keys to unveiling the high-energy theory.

5.4 Electroweak vs. Supersymmetry Breaking

Having established the MSSM matter-matter and matter-gaugino interactions (in the supersymmetric limit in which only D and F terms are considered), the new particle spectrum can be written down. It is instructive to first consider the already available pattern of symmetry breaking in nature, electroweak symmetry breaking, and whether it is sufficient to generate a consistent superpartner spectrum. This is the limit of the MSSM with no explicit supersymmetry breaking. If electroweak $SU(2) \times U(1)$ symmetry is conserved, all particles – fermions and bosons, new and old – are massless in this limit. Electroweak symmetry breaking (EWSB) is responsible for all mass generation and, in particular, is required break the boson - fermion degeneracy. Indeed, supersymmetry is not conserved but is spontaneously broken once electroweak symmetry is broken. As will be shown below, the non-vanishing Higgs *vev's* generate non-vanishing F and D *vev's*, the order parameters of supersymmetry breaking. Let us then postulate for the sake of argument that the Higgs fields acquire *vev's* which spontaneously break $SU(2) \times U(1)$, and consider the impact on supersymmetry breaking on the sparticle spectrum. (As we shall see later, as a matter of fact such *vev's* cannot be generated in this limit.)

Consider, for example, the sfermion spectrum. The spectrum can be organized by flavor (breaking $SU(2)$ doublets to their components), each flavor sector constitute a left-handed $\tilde{f}_L$ and right-handed $\tilde{f}_R$ sfermions (and in principle all sfermions with the same QED and QCD quantum numbers could further mix in a 6×6 subspaces). One has in each sector three possible mass terms $m^2_{LL}\tilde{f}_L\tilde{f}^*_L$, $m^2_{RR}\tilde{f}^*_R\tilde{f}_R$, $m^2_{LR}\tilde{f}_L\tilde{f}^*_R + h.c.$, with $m^2_{LL,RR} \sim \langle D \rangle \sim \nu^2$ generated by substituting the Higgs *vev* in the quartic potential and $m^2_{LR} \sim \langle F \rangle \sim \mu\nu$ generated by substituting the Higgs *vev* in the trilinear potential. The mass matrix for the sfermions $(\tilde{f}_L, \tilde{f}^*_R)$

$$M^2_{\tilde{f}} = \begin{pmatrix} m^2_{LL} & m^2_{LR} \\ m^{2\,*}_{LR} & m^2_{RR} \end{pmatrix}$$

then reads in this limit

$$\left(\begin{array}{cc} m_f^2 + M_Z^2 \cos 2\beta \left[T_3^f - Q_{f_L} \sin^2 \theta_W \right] & m_f \mu^* \tan \beta \, (\text{or} 1/\tan \beta) \\ m_f \mu \tan \beta \, (\text{or} 1/\tan \beta) & m_f^2 - (M_Z^2 \cos 2\beta \times Q_{f_R} \sin^2 \theta_W) \end{array} \right).$$

The sneutrino is an exception since there are no singlet (right-handed) neutrinos and sneutrinos, and its mass is simply $m_{LL}^2 \tilde{\nu}_L \tilde{\nu}_L^*$. Hereafter we will simply write $m_{LR}^2 \tilde{f}_L \tilde{f}_R + h.c.$ with $\tilde{f}_R$ understood to denote the $SU(2)_L$ singlet sfermions (*i.e.*, the conjugation will not be written explicitly).

It is instructive to examine all contributions in some detail (see also exercises). The diagonal fermion mass term is the F-contribution $|y_f \langle H_i^0 \rangle \tilde{f}_{L,R}|^2$. Note that this particular F-term itself has no *vev*, $\langle F \rangle \sim y_f \langle H_i \tilde{f} \rangle = 0$, and hence it does not break supersymmetry but only lead to a degenerate mass contribution to the fermion and to the respective sfermions. The second diagonal contribution is that of the D-term which has a non-vanishing *vev*

$$\langle D \rangle^2 = \frac{g^2 + g'^2}{8} \left(\langle H_2 \rangle^2 - \langle H_1 \rangle^2 \right)^2 = \frac{1}{4} M_Z^2 \nu^2 |\cos 2\beta|^2 = \langle V_{\text{MSSM}} \rangle^2, \quad (5.10)$$

where $\cos 2\beta < 0$ appears only as absolute value (so that $\langle V_{\text{MSSM}} \rangle \geq 0$ as required by (global) supersymmetry); $T_3 = \pm 1/2$ for the Higgs weak isospin was used. Relation (5.10) holds in general (and not only in the supersymmetric limit). The D-term *vev* breaks supersymmetry. However, the limit $\tan \beta \to 1$ corresponds to $\langle D \rangle^2 \to 0$ and supersymmetry is recovered. Indeed in this limit the diagonal sfermion masses squared are given by the respective fermion mass squared. This limit also corresponds to a flat direction in field space (as alluded to in the previous chapter) along which the (tree-level) potential vanishes. A flat direction must correspond to a massless (real) scalar field in the spectrum (its zero mode) so there is a boson whose (tree-level) mass is proportional to the D-term and vanishes as $\cos 2\beta \to 0$. This is the (model-independent) light Higgs boson of supersymmetry. This is a crucial point in the phenomenology of the models which we will return to when discussing the Higgs sector in Chap. 8. It is then straightforward, after rewriting the hypercharge in term of electric charge Q_f, to arrive to the contribution to the sfermion mass squared. One can readily calculate the numerical coefficients $T_3 - Q_f \sin^2 \theta_W$ using $\sin^2 \theta_W \simeq 0.23$ for the weak angle to find $T_3 - Q_f \sin^2 \theta_W \simeq 0.34, 0.14, -0.42, -0.08, -0.27, -0.23, 0.5$ for $f = u_L, u_L^*, d_L, d_L^*, e_L, e_L^*, \nu$ (and their generational replicas).

Lastly, the off-diagonal terms (often referred to as left-right mixing) are supersymmetry breaking terms which split the sfermion spectrum from the fermion spectrum even in the absence of D-term contributions. These terms arise from the cross-terms in F_{H_i}, as do the supersymmetry breaking *vev*'s $\langle F_{H_{1,2}} \rangle^2 = y_f^2 \mu^2 \langle H_{2,1} \rangle^2$. The $\tan \beta$ dependence in mass squared matrix assumes $f = d, l$ ($f = u$), *i.e.*, a superpotential coupling to H_1 (H_2). The presence of the left-right terms implies that the sfermion mass eigenstates have no well-defined chirality association.

Since supersymmetry is spontaneously broken by the Higgs *vev's*, the Higgsinos provide the Goldstino. This is the Achilles heal of this scheme: Let us now define the supertrace function

$$\text{STR}_I \mathcal{F}(O_I) \equiv \sum_I N_{c_I} (-1)^{2S_I} (2S_I + 1) \mathcal{F}(O_I)$$

which sums over any function $\mathcal{F}$ of an object O_I which carries spin S_I. (The summation is also over color and isospin factors which here we pre-summed in the color factor N_c.) In the $\langle D \rangle \to 0$ limit, the supertrace summation over the mass eigenvalues of the fermions and sfermions in each flavor sector is zero! After spontaneous supersymmetry breaking (in this limit) one has $m_{\tilde{f}_{1,2}}^2 = m_f^2 \pm \Delta$ where $\Delta = \langle F \rangle$ is given by the F-*vev* that spontaneously break supersymmetry, in our case the left-right mixing term, so that $\text{STR}M = 0$. Hence, unless $|\langle F \rangle| < m_e \sim 0$, which it cannot be given the lower bound on $|\mu|$ discussed previously, a sfermion must acquire a negative mass squared and consequently the SM does not correspond to a minimum of the potential (and furthermore, QED and QCD could be broken in the vacuum). This situation is, of course, intolerable. It arises because the matter fields couple directly to the Goldstino supermultiplet. (It is a general (supertrace) theorem for any spontaneous braking of global supersymmetry.) Another general implication is that the fermions do not feel the spontaneous supersymmetry breaking (at tree level). Turning on the D-contributions does not improve the situation. Eventhough their sum over a single sector does not vanish, gauge invariance guarantees that their sum over each family vanishes, *i.e.*, negative contributions to some eigenvalues of the mass squared mass, as indeed we saw above.

This comes as no surprise: The Higgs squared mass matrix in this limit simply reads

$$M_H^2 = \begin{pmatrix} \mu^2 & 0 \\ 0 & \mu^2 \end{pmatrix}$$

and has no negative eigenvalues. Thus, it cannot produce the SM Higgs potential and the Higgs fields do no acquire a *vev*, so our *ad hoc* assumption of electroweak symmetry breaking cannot be justified. Note that a negative eigenvalue could arise, however, from an appropriate off-diagonal term. Had we extended the model by including a singlet superfield S, replacing $W = \mu H_1 H_2$ with $W = y_S S H_1 H_2 - \mu_s^2 S$, then electroweak symmetry could be broken while supersymmetry is preserved. One has for the Higgs potential $V = |F_S|^2 = |y_S H_1 H_2 - \mu_s^2|^2$. Its minimization gives $\langle H_1^0 \rangle = \langle H_2^0 \rangle = \mu_S / \sqrt{y_S}$ and $\langle S \rangle = 0$. Electroweak symmetry is broken along the flat direction $V = \langle F_S \rangle^2 = 0$ so that supersymmetry is conserved with $m_{\tilde{f}}^2 = m_f^2$ (and with no left-right mixing!). This is a counter example to our observation that in general electroweak breaking implies supersymmetry breaking. Models in which μ is replaced by a dynamical singlet field are often called the next to minimal extension (NMSSM). The specific NMSSM model described here is, however, still far from leading to a realistic model.

Of course, one needs not go through the exercise of electroweak symmetry breaking in order to convince oneself that the models as they are manifested in this limit are inconsistent. The gluino partner of the gluon, for example, cannot receive its mass from the colorless Higgs bosons and remains massless at tree level, even if electroweak symmetry was successfully broken. It would receive a $\mathcal{O}(\text{GeV})$ or smaller mass from quantum corrections if the winos and bino are massive. Such a light gluino would alter SM QCD predictions at the level of current experimental sensitivity [55] while there is no indication for its existence. (It currently cannot be ruled out in certain mass ranges and if it is the lightest supersymmetric particle [56], in which case it must hadronize.)

We must conclude that a realistic model must contain some other source of supersymmetry breaking. The most straightforward approach is then to parameterize this source in terms of explicit breaking in the low-energy effective Lagrangian.

5.5 Soft Supersymmetry Breaking

Our previous exercises in electroweak and supersymmetry breaking lead to a very concrete "shopping list" of what a realistic model should contain:

- Positive squared masses for the sfermions
- Gaugino masses, particularly for the gluino
- Possibly an off-diagonal Higgs mass term $m^2 H_1 H_2$, as it could lead to the desired negative eigenvalue

We can contrast our "shopping list" with what we have learned we cannot have (in order to preserve the cancellation of quadratic divergences):

- Arbitrary quartic couplings
- Arbitrary trilinear singlet couplings in the scalar potential, $C s \phi \phi^*$
- Arbitrary fermion masses

The two last items are in fact related (and fermion masses are constrained only in models with singlets). Hence, both items do not apply to the MSSM (but apply in the NMSSM). We conclude that acceptable spectrum can be accommodated without altering cancellation of divergences! Such explicit breaking of supersymmetry is soft breaking in the sense that no quadratic divergences are re-introduced above the (explicit) supersymmetry breaking scale.

Let us elaborate and show that sfermion mass parameters indeed are soft. For example, consider the effect of the soft operator $m^2 \phi \phi^*$ and its modification of our previous calculation of the quadratic divergence due to the quartic scalar interaction $y^2 |\phi_1|^2 |\phi_2|^2$:

$$y^2 \int \frac{d^4q}{(2\pi^4)} \frac{1}{q^2} \longrightarrow y^2 \int \frac{d^4q}{(2\pi^4)} \frac{1}{q^2 - m^2} \sim y^2 \int \frac{d^4q}{(2\pi^4)} \left\{ \frac{1}{q^2} + \frac{m^2}{q^4} \right\}$$

$$\sim y^2 \int \frac{d^4q}{(2\pi^4)} \frac{1}{q^2} + i\frac{y^2}{16\pi^2} m^2 \ln \frac{\Lambda_{UV}^2}{m^2}. \tag{5.11}$$

Since the quadratically divergent term is as in the $m = 0$ case, it is still canceled by supersymmetry. Hence, eq. (5.11) translates to a harmless logarithmically divergent mass correction

$$\Delta m_i^2 = -y^2 \frac{m_j^2}{16\pi^2} \ln \frac{\Lambda_{UV}^2}{m^2}, \tag{5.12}$$

and similarly for the g^2 quartic interaction. Note the negative over-all sign of the correction! The logarithmic correction is nothing but the one-loop renormalization of the mass squared parameters due to its Yukawa interactions, and it is negative. The implications are clear - a negative squared mass, for example in the weak-scale Higgs potential, may be a a result of quantum corrections - a subject which deserves a dedicated discussion, and which we will return to in Chap. 7. (Of course, integrating (5.11) properly one finds also finite corrections to the mass parameters which we do not discuss here.)

An important implication of the above discussion is that light particles are protected from corrections due to the decoupling of heavy particles (as heavy as the ultraviolet cut-off scale), as long as the decoupling is within a supersymmetric regime, i.e., $(m_{\text{heavy-boson}}^2 - m_{\text{heavy-fermion}}^2)/(m_{\text{heavy-boson}}^2 + m_{\text{heavy-fermion}}^2) \to 0$. This ensures the sensibility of the discussion of grand-unified theories or any other theories with heavy matter and in which there is tree-level coupling between light and heavy matter (leading to one-loop corrections). Supersymmetry-breaking corrections due to the decoupling of heavy particles are still proportional only to the soft parameters and do not destabilize the theory. (The only caveat being mixing among heavy and light singlets [57].) This persists also at all loop orders.

Our potential $V = V_F + V_D + V_{\text{SSB}}$ now contains superpotential contributions $V_F = |F|^2$, gauge contributions $V_D = |D|^2 +$ gaugino-Yukawa interactions, and contributions that explicitly but softly break supersymmetry, the soft supersymmetry breaking (SSB) terms. The SSB potential most generally consists of

$$V_{\text{SSB}} = m_{j*}^{i\,2}\phi_i\phi^{j*} + \left\{ B^{ij}\phi_i\phi_j + A^{ijk}\phi_i\phi_j\phi_k + C_{i*}^{jk}\phi^{i*}\phi_j\phi_k + \right.$$
$$\left. + \frac{1}{2}M_\alpha\lambda_\alpha\lambda_\alpha + h.c. \right\}. \tag{5.13}$$

The soft parameters were originally classified by Inoue *et al.* [58] and by Girardello and Grisaru [59], while more recent and general discussions include Refs. [60, 61, 62]. Note that the trilinear couplings A and C and the gaugino mass M carry one mass dimension, and also carry phases. The parameters

C (often denoted A') are not soft if the model contains singlets (e.g. in the NMSSM) but are soft otherwise, in particular in the MSSM. In fact it can be shown that they are equivalent to explicit supersymmetry breaking matter fermion masses. Nevertheless, they appear naturally only in special classes of models and will be omitted unless otherwise noted. The scale of the soft supersymmetry breaking parameters $m \sim \sqrt{|B|} \sim |A| \sim |M| \sim m_{\mathrm{SSB}}$ is dictated by the quadratic divergence which is cut off by the mass scale they set (see eq. (5.12)) and hence is given by

$$m_{SSB}^2 \lesssim \frac{16\pi^2}{y_t^2} M_{\mathrm{Weak}}^2 \simeq (1\,\mathrm{TeV})^2. \tag{5.14}$$

The mass scale m_{SSB} plays the role and provides an understanding of the mass scale M_{SUSY} discussed in Sect. 2.3. Ultimately, the explicit breaking is to be understood as the imprints of spontaneous supersymmetry breaking at higher energies rather than be put by hand. (See Chap. 10.) For the purpose of defining the MSSM, however, a general parameterization is sufficient.

Once substituting all flavor indices in the potential (5.13), e.g. $B^{ij}\phi_i\phi_j \to m_3^2 H_1 H_2$ (see Chap. 8) and (suppressing family indices) $A^{ijk}\phi_i\phi_j\phi_k \to A_U H_2 \widetilde{Q}\widetilde{U} + A_D H_1 \widetilde{Q}\widetilde{D} + A_E H_1 \widetilde{L}\widetilde{E}$, one finds that the MSSM contains many more parameters in addition to the 17 free parameters of the SM. The Higgs sector can be shown to still be described by only two free parameters. However, the gauge sector contains three new gaugino mass parameters, which could carry three independent phases; the scalar spectrum is described by five 3×3 hermitian matrices with six independent real parameters and three independent phases each (where we constructed the most general matrix in family space for the Q, U, D, L, and E "flavored" sfermions); and trilinear interactions are described (in family space) by three 3×3 arbitrary matrices A_U, A_D, A_E with nine real parameters and nine phases each (and equivalently for the C matrices). A more careful examination reveals that four phases can be eliminated by field redefinitions and hence are not physical. Hence, the model, setting all $C = 0$ ($A' = 0$) (allowing arbitrary C or A' coefficients) is described by 77 (104) parameters if all phases are zero, and by 122 (176) parameters if phases (and therefore CP violation aside from that encoded in the CKM matrix, often referred to as soft CP violation) are admitted.

5.6 R-Parity and Its Implications

While supersymmetrizing the SM we followed a simple guideline of writing the minimal superpotential that consistently reproduces the SM Lagrangian. Once we realized that supersymmetry must be broken explicitly at the weak scale, we introduced SSB parameters which conserved all of the SM local and global symmetries. While it is the most straightforward procedure, it does not lead to the most general result. In the SM lepton L and baryon B number are accidental symmetries and are preserved to all orders in perturbation theory. (For example, models for baryogenesis at the electroweak phase transition rely on non-perturbative baryon number violation.) Once all fields are elevated to superfields, one can interchange a lepton and Higgs doublets $L \leftrightarrow H_1$ in the superpotential and in the scalar potential, leading to violation of lepton number by one unit by renormalizable operators. The $\Delta L = 1$ superpotential operators, for example, are

$$W_{\Delta L=1} = \pm \mu_{L_i} L_i H_2 + \frac{1}{2} \lambda_{ijk} [L_i, L_j] E_k + \lambda'_{ijk} L_i Q_j D_k, \qquad (5.15)$$

where we noted explicitly the antisymmetric nature of the λ operators. Similarly, one can write SSB $\Delta L = 1$ operators.

Lepton number violation is indeed constrained by experiment, but is allowed at a reasonable level, and in particular, electroweak-scale $\Delta L = 1$ operators could lead to a $\Delta L = 2$ Majorana neutrino spectrum, hence, extending the SM and the MSSM in a desired direction. However, baryon number is also not an accidental symmetry in the MSSM and the

$$W_{\Delta B=1} = \frac{1}{2} \lambda''_{ijk} U_i [D_j, D_k] \qquad (5.16)$$

operators are also allowed by gauge invariance. The combination of lepton and baryon number violation allows for tree-level decay of the proton from $\tilde{s}$ or $\tilde{b}$ exchange $(U)UD \rightarrow (U)\widetilde{D} \rightarrow (U)QL$, i.e., $p \rightarrow \pi l, Kl$. (The squark plays the role of a lepto-quark field in this case and the model is a special case of a scalar lepto-quark theory. See the scalar exchange diagram in Fig. 6.3 below.)

The proton life time is given by

$$\tau_{p \rightarrow e\pi} \sim 10^{-16 \pm 1} \mathrm{yr} (m_{\tilde{f}}/1 \, \mathrm{TeV})^4 (\lambda' \lambda'')^{-2}. \qquad (5.17)$$

It is constrained by the observed proton stability $\tau_{p \rightarrow e\pi} \gtrsim 10^{33}$ yr, leading to the constraint $\lambda' \lambda'' \lesssim 10^{-25}$. The constraint is automatically satisfied if either $\lambda' = 0$ (only B violation) or $\lambda'' = 0$ (only L violation), leaving room for many possible and interesting extensions of the MSSM. (For a review, see Ref. [63], as well as Chap. 11.)

The L and B conserving MSSM, however, is again the most general allowed extension (with minimal matter) if one imposes L and B by hand. It is sufficient for that purpose to postulate a discrete Z_2 (mirror) symmetry, *matter parity*,

$$P_M(\Phi_I) = (-)^{3(B_I - L_I)}. \qquad (5.18)$$

Matter parity is a discrete subgroup of the anomaly free $U(1)_{B-L}$ Abelian symmetry discussed in Chap. 1, and hence can be an exact symmetry if $U(1)_{B-L}$ is gauged at some high energy. Equivalently, one often impose a discrete Z_2 R-parity [64],

$$R_P(O_I) = P_M(O_I) \times (-)^{2S_I} = (-)^{2S_I + 3B_I + L_I}, \qquad (5.19)$$

where O_I is an object with spin S_I. R-parity is a discrete subgroup of a $U(1)_R$ Abelian symmetry under which the supersymmetric coordinate transforms with a charge $R(\theta) = -1$ (which is a conventional normalization), i.e., $\theta \to e^{-i\alpha}\theta$. R-symmetry does not commute with supersymmetry (and it extends its algebra [65]). In particular, it distinguishes the superfield components. Hence, R-parity efficiently separates all SM (or more correctly, its 2HDM type II extension) particles with charge $R_P(O_{SM}) = (+)$ from their superpartners, the sparticles, with charge $R_P(\text{sparticle}) = (-)$, and hence, is more often used. Matter or R-parity correspond to a maximal choice that guarantees stability of the proton. Other choices that do not conserve $B - L$, but only B or L, exist and correspond to Z_3 or higher symmetries (for example, baryon and lepton parities [66, 67] and the θ-parity [68]).

Though the minimal supersymmetric extension is defined by its minimal matter content, it is often defined by also its minimal interaction content, as is the case for R_P invariant models. We will assume for now, as is customary, that R_P is an exact symmetry of the low-energy theory (unless otherwise is stated).

Imposing R-parity on the model dictates to large extent the phenomenology of the model. In particular, the lightest superpartner (LSP) (or equivalently, the lightest $R_P = (-)$ particle) must be stable: It cannot decay to only $R_P = (+)$ ordinary particles since such an interaction would not be invariant under the symmetry. (This is not the case in other examples of symmetries given above which can stabilize the proton while admitting additional superpotential or potential terms.) This is a most important observation with strong implications:

1. The LSP is stable and hence has to be neutral (e.g. a bino, wino, Higgsino, sneutrino, or gravitino) or at least bound to a stable neutral state (as would be the case if the gluino is the LSP and it then hadronizes)
2. Sparticles are produced in the collider in pairs
3. Once a sparticle is produced in the collider, it decays to an odd number of sparticles. In particular, its decay chain must conclude with the LSP. The stable LSP escapes the detector, leading in many cases to a distinctive large missing energy signature. (If a light gravitino is the LSP, for instance, other distinctive signatures such as hard photons exist.)
4. The neutral LSP (in most cases) is a weakly interacting massive particle (WIMP) and it could constitute cold dark matter (CDM) with a sufficient density

The latter point requires some elaboration. Roughly speaking, the relic density of a given dark matter component is proportional to the inverse of its annihilation rate, which in turn is given by a cross section. The cross section is typically proportional to the squared inverse of the exchanged sparticle mass squared. Thus, the relic density is proportional to sparticle mass scale to the fourth power, a relation which leads to model-dependent upper bounds of the order of $\mathcal{O}(\text{TeV})$ on the sparticle mass scale. In practice, the density is a complicated function of various sparticle mass parameters, leading to the relation for the relic density Ωh^2 [69, 70]

$$\Omega_{\text{LSP}} h^2 = \sigma_0 \frac{\left(m_{\tilde{l}}^2 + m_{\text{LSP}}^2\right)^4}{m_{\text{LSP}}^2 \left(m_{\tilde{l}}^4 + m_{\text{LSP}}^4\right)}, \tag{5.20}$$

where $\sigma_0 \approx (460\,\text{GeV})^{-2}/\sqrt{N}$, N here is the number of degrees of freedom at the (cosmological) decoupling of the LSP, and this relation holds in a scenario in which the LSP is the bino which annihilated primarily to leptons l via a t-channel slepton $\tilde{l}$ exchange. (Recall the $\tilde{B}l\tilde{l}^*$ vertex.) Indeed, for $m_{\tilde{l}} \gg m_{\text{LSP}}$ one has $\Omega h^2 \simeq \sigma_0 m_{\tilde{l}}^4/m_{\text{LSP}}^2$. A cautionary note is in place: In certain cases the annihilation process may depend on s-channel resonance annihilation and the resonance enhancement of the cross section may allow for a sufficient annihilation of a WIMP which is a few fold heavier than in the usual case [71]. The upper bound is therefore model dependent.

It should be stressed that the proximity of this upper bound and the one derived from the fine-tuning of the Higgs potential is tantalizing and suggestive. More generally, it relates the CDM to the sparticle mass scale. Particles in various sectors of the theory (which may be linked to the SM sector only gravitationally or by other very weak interactions) could also have masses which are controlled by this SSB scale (for example, by the gravitino mass $m_{3/2}$ discussed in Sect. 4.8): Supergravity interactions can generate such masses in many different sectors. Therefore, CDM candidates may be provided by various sectors of the theory, not necessarily the observable (MS)SM sector. An interesting proposal raised recently is that of an axino dark matter [72], which would weaken any CDM imposed upper-bound on the mass of the ordinary superpartners. (The axino is the fermion partner of the axion of an anomalous Peccei-Quinn or R-symmetry, on which we do no elaborate in these notes.) If, however, one more traditionally chooses to assume that the WIMP is the LSP, collider phenomenology and cosmology need to be confronted [73], providing a future avenue to test such a hypothesis, and furthermore, the nature of the WIMP. (LSP assumptions may also be confronted with a variety of low-energy phenomena, for example, see Ref. [74].)

The collider phenomenology of the models was addressed in Zeppenfeld's [34] Tata's [75, 37] lecture notes. Models of CDM are reviewed, for example, in Refs. [76, 77], and the search for energetic neutrinos from LSP annihilation in the sun was discussed by Halzen [35]. These issues will not be studied here.

5.7 Mass Eigenstates and Experimental Status

We conclude this chapter with a transformation from current to mass eigenstates, which is not a trivial transformation given electroweak symmetry breaking (EWSB): As in our "warm-up" case of a supersymmetric limit (Sect. 5.4), interaction eigenstates with different electroweak charges (and hence, chirality) mix once electroweak symmetry is broken.

The Higgs mass matrix is now

$$M_H^2 = \begin{pmatrix} m_{H_1}^2 + \mu^2 & m_{12}^2 \\ m_{12}^{*\,2} & m_{H_2}^2 + \mu^2 \end{pmatrix}, \tag{5.21}$$

and it has a negative eigenvalue for $m_1^2 m_2^2 < |m_{12}^2|^2$, where $m_i^2 = m_{H_i}^2 + \mu^2$, so that electroweak symmetry can be broken. This condition is automatically satisfied for $m_{H_2}^2 \lesssim -\mu^2$ which often occurs due to negative quantum corrections proportional to the t-quark Yukawa couplings eq. (5.12). (This is the radiative symmetry breaking mechanism, which we will return to in Chap. 7.) One neutral CP odd and two charged $d.o.f.$ are absorbed in the Z and $W^\pm$ gauge bosons, respectively, and two CP even (the lighter h^0 and the heavier H^0), one CP odd (A^0) and one complex charged (H^+) Higgs bosons remain in the physical spectrum. Folding in EWSB constraints, the spectrum is described by only two parameters which are often taken to be $\tan\beta$ and $m_{A^0}^2 = m_1^2 + m_2^2$.

One of the CP even states is the (model-independent) light Higgs boson of supersymmetry which parmeterizes the $\tan\beta = 1$ flat direction mentioned above. This is readily seen in the limit in which all other Higgs $d.o.f.$ form (approximately) a degenerate $SU(2)$ doublet which is heavy with a mass $\sim |\mu| \gg M_W$. In this case, EWSB is SM-like, with the remaining physical CP-even state receiving mass which is proportional to its quartic coupling λ, now given by the gauge couplings $\lambda = (g_2^2 + g'^2)/8 \times \cos^2 2\beta$. In general, its mass $m_{h^0} \leq M_Z |\cos 2\beta| \leq M_Z$ at tree-level, where M_Z reflects the D-term nature of the quartic coupling, and the angular dependence reflects the flat direction. Large $\mathcal{O}(100\%)$ loop corrections, again proportional to the large t-quark Yukawa coupling, lift the (flat direction and the) bound to $m_{h^0} \lesssim \sqrt{2} M_Z \sim 130$ GeV. (Note that perturbation theory does not break down. It is the tree-level term which is small rather than the loop corrections being exceptionally large.) Though the quartic coupling, and hence the mass, can be somewhat larger in extended models, e.g. in some versions of the the NMSSM, as long as perturbativity is maintained one has $m_{h^0} \lesssim 160 - 200$ GeV where the upper range is achieved only in a small class of (somewhat ad hoc) models. The only exception is models with low-energy supersymmetry breaking in which the breaking is not necessarily soft [78]. We discuss the Higgs sector in more detail in Chap. 8 where also references are given.

Sparticles rather than Higgs bosons, however, will provide the evidence for supersymmetry (though the absence of a light Higgs boson can rule most models of perturbative low-energy supersymmetry out). The Dirac-like neutral Higgsinos mix with the other two neutral fermions, the Majorana bino and neutral wino. (The latter can be rewritten as a photino and zino, *i.e.*, as linear combinations aligned with the photon and the Z). The physical eigenstates are the neutralinos $\widetilde{\chi}^0$. Their mass and mixing is given by the diagonalization of the neutralino (tree-level) mass matrix

$$
M_{\widetilde{\chi}^0} = \begin{pmatrix}
M_1 & 0 & -M_Z c_\beta s_W & M_Z s_\beta s_W \\
0 & M_2 & M_Z c_\beta c_W & M_Z s_\beta c_W \\
-M_Z c_\beta s_W & M_Z c_\beta c_W & 0 & \mu \\
-M_Z s_\beta s_W & -M_Z s_\beta c_W & \mu & 0
\end{pmatrix} \quad (5.22)
$$

where M_1 and M_2 are the bino and wino SSB mass parameters, respectively, and $s_\beta = \sin\beta$, $c_\beta = \cos\beta$, and similarly for the weak angle denoted by a subscript W. The neutralino mass matrix is written in the basis $(-i\widetilde{B}, -i\widetilde{W}^0, \widetilde{H}_1^0, \widetilde{H}_2^0)$. The EWSB off-diagonal terms correspond the the gauge-Yukawa interaction terms with the Higgs replaced by its *vev*. Note that in the limit $\mu \to 0$ the wino and bino do not mix at tree-level. Similarly, the charged Higgs and charged gaugino states mix to form the physical mass eigenstates, the charginos $\widetilde{\chi}^\pm$. The chargino mass and mixing is determined by the chargino mass matrix

$$
M_{\widetilde{\chi}^\pm} = \begin{pmatrix}
M_2 & \sqrt{2} M_W s_\beta \\
\sqrt{2} M_W c_\beta & -\mu
\end{pmatrix}, \quad (5.23)
$$

The gluino, of course, cannot mix and has a SSB Majorana mass M_3.

Finally, we can rewrite the sfermion mass matrix for the sfermions $(\widetilde{f}_L, \widetilde{f}_R)$. (Recall that $\widetilde{f}_R$ is a shorthand notation for $\widetilde{f}_R^*$.) The mass-squared matrix

$$
M_{\widetilde{f}}^2 = \begin{pmatrix}
m_{LL}^2 & m_{LR}^2 \\
m_{LR}^{2\,*} & m_{RR}^2
\end{pmatrix} \quad (5.24)
$$

was previously given in the supersymmetric limit. Including the SSB interactions one has

$$
m_{LL}^2 = m_{\widetilde{f}_L}^2 + m_f^2 + M_Z^2 \cos 2\beta \left[T_3^f - Q_{f_L} \sin^2 \theta_W \right], \quad (5.25)
$$

$$
m_{RR}^2 = m_{\widetilde{f}_R}^2 + m_f^2 + M_Z^2 \cos 2\beta \times Q_{f_R} \sin^2 \theta_W, \quad (5.26)
$$

$$
m_{LR}^2 = m_f \left(A_f - \mu^* \tan\beta \right) \ \left[\text{or } m_f \left(A_f - \mu^*/\tan\beta \right) \right], \quad (5.27)
$$

where the first term in (5.25) and in (5.26) is the SSB mass-squared parameter, the triliner SSB parameter is implicitly assumed to be proportional to the Yukawa coupling $\hat{A}_f = y_f A_f$, which is then factored out (the assumption is trivial in the case of one generation but it constitutes a strong constraint in the case of inter-generational mixing), and $\mu \tan \beta$ ($\mu / \tan \beta$) terms appear in the down-squark and slepton mass matrices (up-squark mass matrix), as before.

Neglecting fermion masses, one has the sum rule $m_{\tilde{E}_L}^2 - m_{\tilde{N}_L}^2 = m_{\tilde{D}_L}^2 - m_{\tilde{U}_L}^2 = -\cos^2 \theta_W M_Z^2 \cos 2\beta = -M_W^2 \cos 2\beta > 0$. This sum rule is modified if there is an extended gauge structure with more than just the electroweak D-terms. The A-terms are not invariant under $SU(2) \times U(1)$ and the sfermion doublet masses further split. The mass matrices can be combined to three 6×6 matrices and one 3×3 matrix (for the sneutrinos) just as in the supersymmetric limit discussed in Sect. 5.4. Note that the SSB A-terms $AH\tilde{f}_L\tilde{f}_R$ are holomorphic (and do not involve complex conjugate fields) unlike the supersymmetric trilinear interactions $y\mu H^*\tilde{f}_L\tilde{f}_R$ (and SSB C-terms), a property which is particularly relevant for the stability of the vacuum discussed in Chap. 12. (Note that each holomorphic A-term correpsonds to a flat direction in the scalar potential.) It is also relevant for the dependence of the couplings of the physical eigenstates on $\tan \beta$.

The sfermion left-right mixing angle is conveniently given by

$$\tan 2\theta_{\tilde{f}} = \frac{2m_{LR}^2}{m_{LL}^2 - m_{RR}^2}. \tag{5.28}$$

Obviously, significant left-right mixing is possible if either the corresponding fermion is heavy (as in the case of the stop squarks) or if $\tan \beta$ is large and the corresponding fermion is not very light (as could be the case for the sbottom squarks). Observe, however, that in the limit $m_{\tilde{f}}^2 \gg \langle H_i^0 \rangle^2$ the sfermion mixing is suppressed and mass eigenstates align with the current eigenstates. (This is true assuming $A < m_{RR}^2/\langle H_i^0 \rangle$, $m_{LL}^2/\langle H_i^0 \rangle$, a constraint which is typically impose by the stability of the vacuum. See Chap. 12.)

Model-independent limits (i.e., independent of the decay mode) on the sparticles were given by the total width measurement at the Z pole at LEP. These limits constrain the sparticles to be heavier than $40 - 45$ GeV, with the exception of the lightest neutralino, whose couplings to the Z could be substantially suppressed[2] and could still be as light as ~ 20 GeV. Also, no significant limit on the (heavy) gluino, which does not couple to the Z at tree level, is derived. LEP runs at higher energy further constrain many of the sparticles to be heavier than $\sqrt{s}/2 \sim \mathcal{O}(100)$ GeV. ($\sqrt{s}$ is the experiment center of mass energy, and it is divided by the number of colliding leptons or partons.) However, now the constraints are model-dependent since off-resonance production involves also a t-cahnnel exchange (which introduce strong model dependence) and, in the absence of an universal tool such as the Z width, searches must assume in advance the decay chain and its final

---

[2] One could also fine-tune the $\sin \theta_{\tilde{f}} / \sin \theta_W$ ratio so that a particular sfermion $\tilde{f}_i$ decouples from the Z. Such tuning, however, is not scale invariant.

products. The FNAL Tevatron can constrain efficently the strongly inter-
acting squarks and gluino with lower bounds of $\sqrt{s}/6 \sim \mathcal{O}(200\text{-}300)$ GeV,
but again, these bounds contain model-dependent assumptions. Many spe-
cialized searches assuming unconventional decay chains were conducted in
recent years both at the Tevatron and LEP. They are summarized and corresponding limits are updated periodically by the Particle Data Group [7].
Obviously, a significant gap remains between current limits (and particularly
so, the model-independent limits) and the theoretically suggested range of
$\mathcal{O}(1\,\text{TeV})$. Though some sparticles may still be discovered at future Tevatron runs, it is the Large Hadron Collider, currently under consttruction at
CERN, which will explore the $\mathcal{O}(1\,\text{TeV})$ regime and the "TeV World".

Exercises

5.1 Draw all the Feynamn diagrams that stem from that of the vector-
fermion-fermion interaction and which describe the gauge-matter interactions. What are the Feynman rules for the case of a $U(1)$?

5.2 Confirm that $\mathrm{Tr}Y^3 = 0$ for hypercharge in the MSSM, as well as the
mixed anomaly traces $\mathrm{Tr}SU(N)^2U(1)$ where $N = 2, 3$.

5.3 Derive the complete MSSM component-field Yukawa Lagrangian.

5.4 Derive the complete MSSM quartic potential, including gauge (D) and
Yukawa (F) contributions. Use the relation $\sigma_{ij}^a\sigma_{kl}^a = 2\delta_{il}\delta_{jk} - \delta_{ij}\delta_{kl}$ among
the $SU(2)$ generators to write the Higgs quartic potential explicitly. Map it
onto the general form of a 2HDM potential eq. (3.3) and show that (at tree
level) $\lambda_1 = \lambda_2 = (1/8)(g'^2 + g_2^2)$; $\lambda_3 = -(1/4)(g'^2 - g_2^2)$; $\lambda_4 = -(1/2)g_2^2$; and
$\lambda_5 = \lambda_6 = \lambda_7 = 0$ (where g' and g_2 are the hypercharge and $SU(2)$ couplings,
respectively).

5.5 Show that at the quantum level there could be non-diagonal corrections to the gauge kinetic function $f_{\alpha\beta}$, for example, in a theory involving
$U(1) \times U(1)'$ [79]. (Consider a loop correction that mixes the gauge boson
propagators.)

5.6 Derive the D-term expectation value eq. (5.10). Rewrite the hypercharg
in terms of the electric charge, weak isospin, and the weak angle to derive
the D term contribution to the sfermion mass squared. Confirm its flavor-
dependent numerical coefficient.

5.7 Derive the Higgs mass-squared matrix in the supersymmetric limit.

5.8 Derive the Higgs mass-squared matrix in the model with a singlet S
which is described in Sect. 5.4. Show that it has an off-diagonal element and
that it has a negative eigenvalue. Show, by considering F_{H_i} contributions to
the scalar potential, that indeed $\langle S \rangle = 0$ and no left-right sfermion mixing
arises.

5.9 Show that all the soft parameters are indeed soft by naive counting, where possible (e.g. in the case of trilinear scalar interactions) or otherwise by integration.

5.10 Show that a supersymmetry breaking fermion mass, if allowed by the gauge symmetries (for example, consider a toy model with two singlets $W = S_1 S_2^2 + m S_1^2$, and a supersymmetry breaking mass to the fermion component of S_2 $V_{\rm SSB} = \widetilde{\mu} \psi_2 \psi_2$) can be recast, after appropriate redefinitions, as a $C s_1 s_2 s_2^*$ term in the scalar potential (with $C = -\widetilde{\mu}$).

5.11 Show that in a model with a singlet (consider, for example, our toy model above) an interaction $C s \phi_1 \phi_2^*$ is not soft but leads to a quadratically divergent linear term in the scalar potential. Return to Ex. 5.10 and write the tadpole diagram which leads to a quadratically-divergent linear term for the singlet s_1 in either langauge, and confirm their equivalence.

5.12 Substitute the SM flavor indices in the soft potential (5.13) and confirm our counting of free parameters.

5.13 Write the most general gauge invariant (R-parity violating) SSB potential. How many parameters describe the MSSM if R-parity is not imposed?

5.14 Using the (continuous) R-charge assignment of the coordinate θ, what are the R-charges of the various component fields of the chiral and vector superfields? and of the superpotential? Show that requiring that the potential is R-invariant forbids gaugino masses and trilinear scalar terms. Therefore, these parameters must carry R-charge and their presence breaks the $U(1)_R$ symmetry. Note that all of the above terms are invariant under the discrete R_P subgroup!

5.15 Calculate the upper bound on the slepton mass implied by the approximation eq. 5.20 for $\Omega_{\rm LSP} h^2 \leq 1$ and $m_{\rm LSP} = 100$ GeV.

5.16 Diagonalize the neutralino and chargino mass matrices in the limits $\mu \to \infty$ (gaugino region), $M_1 \sim M_2 \to \infty$ (Higgsino region), and $\mu \sim M_1 \sim M_2 \to 0$.

5.17 Extend the neutralino and chargino mass matrices to include mixing with one generation of neutrinos and charged leptons, respectively, for $\mu_{L_3} \neq 0$. The neutrino mass is given by the ratio of the determinant of this matrix and that of the usual neutralino mass matrix. Can you identify the limit in which one mass eigenvalue (the neutrino mass) is zero (at tree-level)? What is the effective LSP in these models? Allow neutralino mixing with three generations of neutrinos $\mu_{L_a} \neq 0$ for $a = 1, 2, 3$. Show that still only one neutrino could be massive at tree level.

5.18 Show that each holomorphic A-term corresponds to a flat D-term direction. Estimate the upper bound on A by considering tachion states in the sfermion mass matrix.

5.19 (a) In the MSSM, like in the SM, the decay $h^0 \to b\bar{b}$ is often the dominant decay mode of the light CP-even Higgs boson. Nevertheless, other decay channels may also be open, e.g. $h^0 \to \gamma\gamma$ plays an important role is search strategy at the LHC. Write the Feynman diagram for a supersymmetric invisible decay mode of h^0, both assuming R-parity conservation and violation. The stable or meta-stable invisible states escape the detector.

5.19 (b) In some models the Goldstino is the LSP, and the next to lightest supersymmetric particle (NLSP) is charged and stable on collider scales, for example the stau. What would be a potential signature of such a scenario?

Summary

In this part of the notes supersymmetry was motivated, constructed, and applied to the Standard model of particle physics. The minimal model was defined according to its particle content and superpotential, but it was shown to contain many arbitrary parameters that explicitly break supersymmetry near the Fermi scale. Indeed, many questions still remain unanswered:

- Does the model remain perturbative and if so, up to what scale?
- What is the high-energy scale which we keep referring to as the ultraviolet cut-off scale?
- What is the origin of the soft supersymmetry breaking parameters and can the number of free parameters be reduced?
- Are there signatures of supersymmetry (short of sparticle discovery) that could have been tested at past and current low-energy experiments?
- Could one distinguish a MSSM Higgs boson from a SM one?
- Can the model be extended to incorporate neutrino mass and mixing?

In addition, one would like to understand how all the different aspects of supersymmetry – perturbativity, renormalization, ultraviolet origins, etc. come together to explain the weak scale and its structure. This and related issues have been extensively studied in recent years. While some possible answers and proposals were put forward, no standard *high-energy* supersymmetric model exists. On the contrary, the challenge ahead is the deciphering of the high-energy theory from the low-energy data once supersymmetry is discovered and established.

We will address these and similar questions in the remaining parts of these notes. In particular, we will try to link the infrared and the ultraviolet.

Part III

Supersymmetry Top-Down:
Understanding the Weak Scale

6. Unification

Shortly after the SM was established and asymptotic freedom realized, it was suggested that the SM semi-simple product group of $SU(3) \times SU(2) \times U(1)$ is embedded at some high energies in a unique simple group, for example $SU(5)$ (which like the SM group has rank 5 and hence is the smallest simple group that can contain the SM gauge structure). Aside from implications for quark-lepton unification, Higgs fields, and the proton stability (to which we will return) it predicts first and foremost that the seemingly independent SM gauge couplings originate from a single coupling of the unified simple group [80] and that their infrared splitting is therefore due to only the scaling (or renormalization) of the corresponding quantum field theory from high to low energies [81]. Independently, it was also realized in the context of string theory that the theory just below the string scale often (but not always) has a unique (or unified) value for all gauge couplings [19, 18, 82, 83], regardless of whether the SM group itself truly unifies (in the sense of (i) embedding all SM fields in representations of some simple group, and (ii) spontaneous breaking of that group).

The question is then obvious: Do the measured low-energy couplings unify (after appropriate scaling, i.e., renormalization) at some ultraviolet energies? Clearly, if this is the case then their unification automatically defines an ultraviolet scale (which is a reasonable choice to many of the exercises that we will undertake in the following chapters), and supergravity may further facilitate gauge-gravity unification. This question is readily addressed using the renormalization group formalism (which is beautifully confirmed by the data in the case of QCD at energies up to the LEP center-of-mass energy of ~ 200 GeV). Once the particle content of the model is specified, the β function coefficients b can be calculated and the couplings can be extrapolated by integrating the renormaization group equation $dg/d\ln \Lambda = (b/16\pi^2)g^3$ (given here at one loop). Such an integration is nothing but the scaling of the theory, e.g. between the weak scale, where the couplings are precisely known, and some high-energy scale, assuming a specific particle content. Such a scaling can be performed in any perturbative theory as long as the particle content is known or fixed. (For example, the SM scaling of QCD below the weak scale agrees with the measurement of the QCD coupling at various energies. See Fig. 6.1.) Its conclusions are meaningful if and only if the low-energy couplings

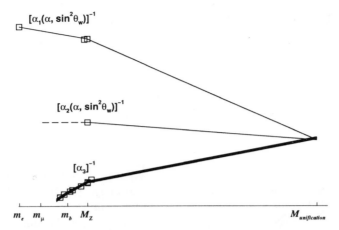

Fig. 6.1. The SM gauge couplings evolve from their measured low-energy value (indicated by the square symbols) to high energies using SM three-loop (for scales $\Lambda < M_Z$) and MSSM two-loop (for scales $\Lambda > M_Z$) β-functions. Their renormalization within the SM is confirmed by the data, while extrapolation to higher energies, assuming the MSSM, leads to their unification at nearly a point. Note that (only) α_3 exhibits asymptotically-free behavior.

are measured to such precision so that their experimental errors allow only a sufficiently small range of values for each of the extrapolated high-energy couplings. In particular, one needs to be able to determine whether all three coupling intersects to a satisfactory precision at a point, and if perturbativity (which is used in deriving the equations) is maintained.

The 1990's brought about the most precise determination of the gauge couplings at the Z pole (see summary by Langacker [84]), therefore enabling one to address the question. Assuming only the SM matter content at all energies, the three couplings fail to unify – their integration curves $g(\ln \Lambda)$ intersects in pairs only. (Of course, one may question the validity of such a framework to begin with in the absence of any understanding as to how is the hierarchy problem resolved in this case.) Extending the SM to the MSSM and using the MSSM particle content begining at some point between the weak scale and a TeV scale, the three SM gauge coupling unify at a point at a scale $M_U \simeq 3 \times 10^{16}$ GeV and with a value $g_U \simeq 0.7$ or $\alpha_U \simeq 0.04$, defining the unification scale (which is then a potential candidate for providing the ultraviolet cut off) as well as confirming perturbativity, and hence consistency, of the framework. This is illustrated in Fig. 6.1. See also Ref. [85].

6.1 Gauge Coupling Unification

Let us repeat this exercise in some detail. The renormalization of the couplings is given at two-loop order by

$$\frac{dg_i}{\ln \Lambda} = \frac{b_i}{16\pi^2}g_i^3 + \frac{b_{ij}}{(16\pi^2)^2}g_i^3 g_j^2 - \frac{a_{i\alpha}}{(16\pi^2)^2}g_i^3 y_\alpha^2. \tag{6.1}$$

The first term is the one-loop term given above, while the second and third terms sum gauge and Yukawa two-loop corrections, respectively, and we did not include any higher loop terms which are negligible in the MSSM. (The two-loop equation is free of any scheme dependences, though negligible (finite) corrections appear when matching the scheme used to describe the data to a scheme which can be used for extrapolation in supersymmetry.) The Yukawa terms, which are negative ($a_{i\alpha} > 0$), correspond to only small $\mathcal{O}(1\%)$ but model ($\tan\beta$-) dependent corrections, which we again neglect here. The gauge two-loop terms correspond to $\mathcal{O}(10\%)$ corrections, which we will include in the results. The one loop coefficients are the most important. In supersymmetry they are conveniently written as a function of the chiral superfields quantum numbers,

$$b_i = \sum_a T_i(\Phi_a) - 3C_i, \tag{6.2}$$

where the Dynkin index $T_i(\Phi_a) = 1/2\,(Q^2)$ for a superfield Φ_a in the fundamental representation of $SU(N)$ (with a $U(1)$ charge Q), and the Casimir coefficient for the adjoint representation $C_i = N\,(0)$ for $SU(N)\,(U(1))$. It is now straightforward to find in the MSSM

$$\begin{pmatrix} b_1 \\ b_2 \\ b_3 \end{pmatrix} = N_{\text{family}} \begin{pmatrix} 2 \\ 2 \\ 2 \end{pmatrix} + N_{\text{Higgs}} \begin{pmatrix} \frac{6}{10} \\ 1 \\ 0 \end{pmatrix} - \begin{pmatrix} 0 \\ 6 \\ 9 \end{pmatrix} = \begin{pmatrix} +\frac{66}{10} \\ +1 \\ -3 \end{pmatrix}, \tag{6.3}$$

where N_{family} and N_{Higgs} correspond to the number of chiral families and Higgs doublet pairs, respectively, and the index $i = 1, 2, 3$ correspond to the $U(1)$, $SU(2)$, and $SU(3)$, coefficients.

Here, we chose the "unification" normalization of the hypercharge $U(1)$ factor, $Y \to \sqrt{3/5}Y$ and $g' \to \sqrt{5/3} \equiv g_1$. This is the correct normalization if each chiral family is to be embedded in a representation(s) of the unified group. In this case, the trace over a family over the generators T^α for each of the SM groups i, $\text{Tr}(T_i^\alpha(\Phi_a)T_i^\beta(\Phi_a)) = n\delta^{\alpha\beta}$, has to be equal to the same trace but when taken over the unified group generators. Reversing the argument, the trace has therefore to be equal to the same n for all subgroups. In the MSSM, both non-Abelian factors have $n = 2$ and can indeed unify. The $U(1)$ normalization is chosen accordingly such that $\text{Tr}(Q_a^2) = 2$ for each family. On the same footing, this condition and eq. (6.2) imply that adding any complete multiplet(s) (e.g. a SM family) of a unified group modifies b_1, b_2, and b_3 exactly by the same amount. (Note, however, that asymptotic freedom of QCD may be lost if adding such multiplets. $SU(2)$ is already not asymptotically free in the MSSM.) This relation also explains, in the context of unification, the quantization of hypercharge which is dictated by the embedding of the $U(1)$ in a non-Abelian group, and by the breaking pattern of the higher rank group. (Such an embedding of the SM group in a single non-Abelian group implies an anomaly free theory, and hence, this relation is somewhat equivalent to the low-energy quantization based on the anomaly constraints.)

Neglecting the Yukawa term, the coupled two-loop equations can be solved in iterations,

$$\frac{1}{\alpha_i(M_Z)} = \frac{1}{\alpha_U} + b_i t + \frac{1}{4\pi} \sum_{j=1}^{3} \frac{b_{ij}}{b_j}(1 + b_j \alpha_U t) - \Delta_i, \qquad (6.4)$$

where the integration time is conveniently defined $t \equiv (1/2\pi)\ln(M_U/M_Z)$, and we assume, for simplicity, that the MSSM β-functions can be used from the Z scale and on. The threshold function Δ_i compensates for this naive assumption and takes into account the actual sparticle spectrum. It can also contain contributions from additional super-heavy particles and from operators (e.g., non-universal corrections to $f^i_{\alpha\beta}$) that may appear near the unification scale.

These equations can be recast in terms of the fine-structure constant $\alpha(M_Z)$ and the weak angle $\sin^2 \theta_W \equiv s_W^2(M_Z)$. Using their precise experimental values one can then calculate (up to threshold corrections) the unification scale, the value of the unified coupling, and the value of the low-energy strong coupling. (Since there are only two high-energy parameters, one equation can be used to predict one weak-scale coupling.) The first two are predictions that only test the perturbativity and consistency of the extrapolation. (Note that the unification scale is sufficiently below the Planck scale so that gravitational correction may only constitute a small perturbation which can be summed, in principle, in Δ_i.)

One finds, given current values,

$$\alpha_3(M_Z) = \frac{5(b_1 - b_2)\alpha(M_Z)}{(5b_1 + 3b_2 - 8b_3)s_W^2(M_Z) - 3(b_2 - b_3)} + \cdots \qquad (6.5)$$

$$= \frac{7\alpha(M_Z)}{15s_W^2(M_Z) - 3} + \text{two-loop and threshold corrections}$$

$$= 0.116 + 0.014 - (0.000 - 0.003) \pm \delta$$

$$= (0.127 - 0.130) \pm \delta, \qquad (6.6)$$

where in the third line one-loop, two-loop gauge, two-loop Yukawa, and threshold corrections (denoted by δ) are listed. Note that any additional complete multiplets of the unified group in the low-energy spectrum would shift all the b_i's by the same amount and therefore factor out from the (one-loop) prediction. This is true only at one loop and only for the $\alpha_3(M_Z)$ and t predictions.

Comparing the predicted value to the experimental one (whose precision increased dramatically in recent years) $\alpha_3(M_Z) = 0.118 \pm 0.003$, one finds a $\sim 8\%$ discrepancy, which determines the role that any structure near the weak scale, the unification/gravity scale, or intermediate scales, can play. In fact, only $\sim 3\%$ corrections are allowed at the unification scale, since the QCD renormalization amplifies the corrections to the required 8% at the electroweak scale. GUT-scale corrections are the most likely conclusion from the discrepancy, but we will not discuss here in detail the many possible sources of such a small perturbation which could appear in the form of a super-heavy non-degenerate spectrum, non-universal corrections to the gauge kinetic function once the grand-unified theory is integrated out, string corrections (in the case of a string interpretation), etc. Let us instead comment on the possible threshold structure at the weak scale (ignoring the possibility of additional complete multiplets at intermediate energies which will modify the two-loop correction [86]).

Unless some particles are within tens of GeV from the weak scale, only the (leading-)logarithm corrections need to be considered (either by direct calculations or using the renormalization group formalism). This is only the statement that the sparticle spectrum is typically expected to preserve the custodial $SU(2)$ symmetry of the SM and hence not to contribute to the (universal) oblique (e.g. ρ) and other (non-universal) parameters that measure its breaking and which would contribute non-logarithmic corrections to δ.

The leading logarithm threshold corrections can be summed in a straightforward but far from intuitive way in the threshold parameter [85] $\hat{M}_{SUSY}$,

$$\delta = -\frac{19}{28}\frac{\alpha_3^2(M_Z)}{\pi}\ln\frac{\hat{M}_{SUSY}}{M_Z} \simeq -0.003\ln\frac{\hat{M}_{SUSY}}{M_Z}. \tag{6.7}$$

If all sparticles are degenerate and heavy with a mass $\hat{M}_{SUSY} \simeq 3$ TeV then the predicted and experimental values of the strong coupling are the same. However, more careful examination shows that $\hat{M}_{SUSY}$ is more closely related to the gaugino and Higgsino spectrum (since the sfermion families correspond to complete multiplets of $SU(5)$) and furthermore, typical models for the spectrum give $10 \lesssim \hat{M}_{SUSY} \lesssim 300$ GeV even though the sparticles themselves are in the hundreds of GeV range. The threshold correction may then be even positive! (This is particularly true when non-logarithmic corrections are included.) While high-energy contributions to δ are likely to resolve the discrepancy, they also render the unification result insensitive to the exact value of $\hat{M}_{SUSY}$, since, e.g. an additional deviation of $\sim 3\%$ from $\hat{M}_{SUSY} \ll M_Z$ corresponds to only $\sim 1\%$ corrections at the unification scale. Hence, even though the sparticle threshold corrections may not resolve the discrepancy, they are very unlikely destroy the successful unification.

This low level of ultraviolet sensitivity (or rather high insensitivity) more then allows one to trust the result. (If the factor of $2\pi t \simeq 30$ corresponds to a power-law (rather than logarithmic) renormalizations, as in some model in which the gauge theory is embedded in a theory with intermediate-energy extra dimensions [22], the sensitivity is amplified by more than an order of magnitude, undermining any predictive power in that case.) The predictive power in the minimal framework is unique and one of the pillars in its support.

One could then take the point of view that the SM is to be embedded at a grand-unification scale $M_U \simeq 3 \times 10^{16}$ GeV in a simple grand-unified (GUT) group, $SU(3)_c \times SU(2)_L \times U(1)_Y \subseteq SU(5); SO(10); SU(6); E_6 \cdots$ where the rank five and six options were specified. The GUT group may or may not be further embedded in a string theory, for example. Alternatively, the unification scale may be interpreted as a direct measurement of the string scale. The latter interpretation is one of the forces that brought about a revolution in string model building, which generically had the string scale an order of magnitude higher, and motivated M-theory model building and other non-traditional approaches [18], which were reviewed recently by Dienes [19]. In particular, the 11-dimensional M-theory, with a fundamental scale $M_{11} \gtrsim M_U$, allows a finite and small (six-dimensional) compactification volume $V \sim M_U^{-6}$ and a finite and small separation R_{11} between the six- and four- dimensional "walls" (see Sect. 2.2.2) such that $\alpha_U \sim M_{11}^{-6}/V \sim (M_U/M_{11})^6$ can be adjusted to fit its "measured" value of ~ 0.04 [18].

A GUT group with three chiral generations; and the appropriate Higgs representations (that can break it to the SM group as well as provide the SM Higgs doublets) can be further embedded, in principle, in a string theory a decade or so above the unification scale. However, this idea encounters many difficulties and was not yet demonstrated to a satisfactory level [19]. In the reminder of our discussion of unification we will outline the embedding in a GUT group, leaving open the question of string theory embedding.

6.2 Grand Unification

Grand-unification, in contrast to only coupling unification, requires one to embed all matter and gauge fields in representations of the large GUT group. For example, all the SM gauge (super)fields are embedded in the case of $SU(5)$ ($SO(10)$) in the **24** (**45**) dimensional adjoint representation, while each family can be embedded in $\{D, L \subseteq \bar{\mathbf{5}}\} + \{Q, U, E \subseteq \mathbf{10}\}$ ($\{Q, U, D, L, E, N \subseteq \mathbf{16}\}$) anti-fundamental + antisymmetric (spinor) representations. The Higgs fields of the SM can be embedded in $\bar{\mathbf{5}} + \mathbf{5}$ (one or more fundamental **10**'s and anti-fundamentals). The Higgs sector must also contain a higher dimensional representation(s) (which may be taken to be the adjoint) that can break the

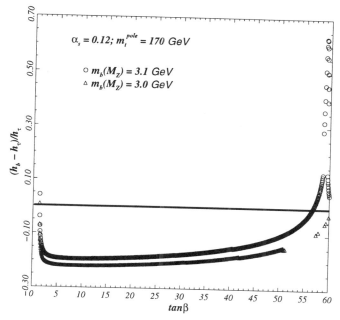

Fig. 6.2. The unification-scale difference $y_b - y_\tau$ (denoted here $h_b - h_\tau$) is shown in y_τ (denoted here h_τ) units for $m_b(M_Z) = 3$ GeV, $\alpha_3(M_Z) = \alpha_s(M_Z) = 0.12$, $m_t^{pole} = 170$ GeV and as a function of $\tan\beta$. The zero line corresponds to $b - \tau$ unification. For comparison, we also show the difference for $m_b(M_Z) = 3.1$ GeV (which for $\alpha_s(M_Z) = 0.12$ is inconsistent with $m_b(m_b) < 4.45$ GeV). Note the rapid change near the (naive) small and large $\tan\beta$ solutions, which is a measure of the required tuning in the absence of threshold corrections. Also note that in most of the parameter space the unification-scale leptonic coupling is the larger coupling. Taken from Ref. [90].

GUT symmetry spontaneously to its SM subgroup, and whose interactions must be described by the appropriate superpotential.

Note that the GUT theory resides in the globally supersymmetric SM sector with all minima corresponding to $\langle V \rangle = 0$. Therefore, the GUT $\rightarrow$ SM minimum is at best degenerate with the GUT conserving or any other minimum. It is the supergravity effects, which typically can be parameterized by the (small) soft parameters, that must lift the degeneracy and pick the correct SM minimum.

Various issues stem from the embedding, and we will touch upon the most important ones next. Recommended readings include Ref. [85], which lists many of the earlier works. A sample of other recent research papers is given in Ref. [87]. Grand-unified theories were reviewed by Mohapatra [88].

6.2.1 Yukawa Unification

First and foremost, quarks and leptons unify in the sense that they are embedded in the same GUT representations (e.g. the down singlet and the lepton doublet are embedded in the $\bar{5}$ of $SU(5)$). This translates to simple boundary conditions for ratios of their Yukawa couplings, e.g. $y_d/y_l = 1$ at the unification scale [89]. This relation (applied to a given generation) can then be renormalized down to the weak scale,

$$
\frac{d}{d\ln\Lambda}\left(\frac{y_d}{y_l}\right) = \frac{1}{16\pi^2}\left(\frac{y_d}{y_l}\right)\left\{y_u^2 + 3(y_d^2 - y_l^2) - \frac{16}{3}g_3^2 - \frac{4}{3}g_1^2\right\}, \qquad (6.8)
$$

and tested. This is shown in Fig. 6.2 taken from Ref. [90]. (Note that the fermion masses are the current masses evaluated at M_Z where $m_b \simeq 3$ GeV and $m_\tau \simeq 1.7$ GeV.) Indeed, it is found to be correct for the third family couplings ($b - \tau$ unification) for either $\tan\beta \simeq 1 - 2$ (large top Yukawa coupling $y_t(M_Z) \simeq 0.95/\sin\beta$) or $\tan\beta \gtrsim 50$ (large bottom Yukawa coupling $y_b(M_Z) \simeq 0.017\tan\beta$), and for a much larger parameter space once finite corrections to the quark masses from sparticle loops (not included here) are considered, e.g. $|\Delta(m_b)/m_b| \lesssim 2\%\tan\beta$.

Clearly, the successful renormalization of the unification relation $y_d/y_l = 1$ requires large Yukawa couplings (which renormalize y_d). The large Yukawa couplings are needed to counterbalance the QCD corrections. Henceforth, it is not surprising that these relations fail for the lighter families: Yukawa unification (in its straightforward form) applies to and distinguishes the third family. The first and second generation fermion masses may be assumed to vanish at the leading order,

$$
Y_{f_{ab}} \simeq y_{f_3}\begin{pmatrix} 0 & 0 & 0 \\ 0 & 0 & 0 \\ 0 & 0 & 1 \end{pmatrix},
$$

and could be further understood as setting the magnitude of the perturbations for any such relations (either from higher dimensional Higgs representations or from Planck-mass suppressed operators, or both).

6.2.2 Proton Decay

Secondly, additional non-SM matter and gauge (super)fields appear, and in most cases must be rendered heavy. For example, in $SO(10)$ the $\mathbf{16}$ contains also a singlet right-handed neutrino N, which is useful for understanding neutrino masses. (We return to neutrino masses in Chap. 11.)

Other heavy particles, however, are more troublesome. The Higgs doublets are embedded together with color triplets which interact with SM matter with the same Yukawa couplings as the Higgs doublets. E.g. the matter $\mathbf{10}_a + \bar{\mathbf{5}}_a$ superpotential in minimal $SU(5)$ (recall $y_d = y_l$ in the minimal

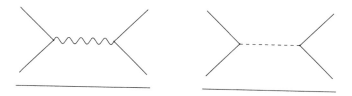

Fig. 6.3. Tree-level proton decay via "lepto-quark" gauge boson and scalar exchange $qq' \to q''l$. The straight line represents the spectator quark which is contained in proton and after its decay hadronizes with q'' to form a meson. q (l) denotes a generic quark (lepton).

model) reads $W \sim y_{u_{ab}} 5_H 10_a 10_b + y_{d_{ab}} \bar{5}_H 10_a \bar{5}_b$, and the 5_H $(\bar{5}_H)$ contains the H_2 (H_1) Higgs doublet and a down-type color triplet H_2^C (H_1^C) with a hypercharge $-1/3$ $(+1/3)$. Also, the **24** contains not only the SM gauge content but also the $(3, 2)_{-5/6}$ $(\bar{X}, \bar{Y})$ vector superfields, as well as the complex conjugate representation (X, Y), all of which must become massive after the spontaneous (GUT) symmetry breaking.

The X and Y are lepto-quark (super)fields which connect lepton and quark fields and therefore lead to proton decay, e.g. $p \to \pi^0 e^+$, at tree level. So do the color triplet Higgs fields, which therefore must also become massive with a mass near the unification scale. (Relevant diagrams are illustrated in Fig. 6.3.) The constraint from the proton life time measurement,

$$\tau_P \simeq 3 \times 10^{31 \pm 1} \text{ yr } \left(\frac{M_U}{4.6 \times 10^{14} \text{ GeV}} \right)^4 \gtrsim 10^{33} \text{ yr}, \tag{6.9}$$

is, however, easily satisfied for $M_U \simeq 3 \times 10^{16}$ GeV.

Nevertheless, supersymmetry implies also heavy color triplet Higgsinos $\widetilde{H}_i^C$ with lepto-quark intergenerational Yukawa couplings, for example, $y_{ab} \widetilde{H}_1^C Q_a L_b$ (derived from $W \sim y_{d_{ab}} \bar{5}_H 10_a \bar{5}_b$). These terms induce proton decay $p \to K^+ \bar{\nu}$ radiatively from loop (box) diagrams[1] with the colored Higgsino and the MSSM gauginos and sfermions circulating in the loop. The colored Higgsino interaction in this diagrams can be described by the effective dimension five operators in the superpotential $W \sim (QQQL + UDUE)/M_U$ whose F component leads to dimension-five vertices of the form $\tilde{f}\tilde{f}ff/M_U$. In this alternative description the diagram can be cast as a triangle sparticle loop: The sparticle dressing loop. In order for the first operator not to vanish one Q field must carry a different generation label, hence, $p \to Kl$. (See exercise.)

---

[1] These diagrams are similar to the meson mixing diagram Fig. 9.1 only with fermion Majorana (gaugino) and Dirac (colored Higgsino) mass insertions instead of the sfermion mass insertions.

Such diagrams $\propto (g^2 y_{ab}^2/16\pi^2)(1/\widetilde{m}M_U)$, where $\widetilde{m} \sim M_{\text{Weak}}$ is a typical superpartner mass scale. In comparison the tree-level diagrams Fig. 6.3 $\propto (g^2/M_U^2)$, (y_{ab}^2/M_U^2). Indeed, radiative decay proved to generically dominate proton decay in these models since the corresponding amplitude is suppressed by only two (rather than four) powers of the superheavy mass scale. (The additional suppression by small intergenerational Yukawa couplings is therefore crucial.) It is interesting to note that these operators are forbidden if instead of R-parity a discrete Z_3 baryon parity is imposed, which truly conserve baryon number but allows R-parity lepton number violating operators [66].

The radiative proton decay places one of the most severe constraints on the models and eventually will determine their fortune. This is particularly true after significant improvements in experimental constraints from the Super-Kamiokande collaboration [91] which yield comparable life-time bounds $\sim \mathcal{O}(10^{33})$ yr on the $K\nu$ and πe decay modes. On the other hand, the large predicted (in relative terms) amplitudes for radiative proton decay provide an opportunity to test the unification framework and supersymmetry simultanoeusly.

6.2.3 Doublet–Triplet Splitting

Our discussion above leads to the other most difficult problem facing the grand-unification framework. Supersymmetry guarantees that once the Higgs doublets and color triplets are split so that the former are light and the latter are heavy, this hierarchy is preserved to all orders in perturbation theory. Nevertheless, it does not specify how such a split may occur. This is the doublet-triplet splitting problem which is conceptually, though (because of supersymmetry non-renormalization theorems) not technically, a manifestation of the hierarchy problem.

More generally, it is an aspect of the problem of fixing the μ-parameter (i.e., the doublet mass) $\mu = \mathcal{O}(M_{\text{Weak}})$, which was mentioned briefly in the previous chapter. Like all other issues raised, extensive model-building efforts and many innovative solutions exist [88], but will not be reviewed here. They typically involve extending the model representations, symmetries and/or (unification-scale) space-time dimensions.

It is intriguing, however, that the fundamental problems of this framework, the light fermion spectrum (e.g. y_{ab}) and the doublet-triplet splitting (e.g. the colored Higgsino mass) combine to determine the proton decay amplitude $p \to K\nu$, which in turn provides a crucial test of the framework, and consequently its potential downfall.

Exercises

6.1 Calculate the β-function coefficients in the MSSM (eq. (6.3)). At what order the assumption of R_P conservation affects the calculation?

6.2 Confirm the hypercharge $U(1)$ GUT normalization.

6.3 Solve in iterations the two-loop renormalization group equation for the gauge coupling, neglecting Yukawa couplings. Rewrite the solutions as predictions for t, M_U and $\alpha_3(M_Z)$.

6.4 Use the strong coupling to predict the weak angle. Show that at the unification scale one has the boundary condition $s_W^2(M_U) = g'^2/(g'^2+g_2^2)|_{M_U} = 3/8$. Compare your prediction to the data.

6.5 Count degrees of freedom and show that the **16** of $SO(10)$ contains a singlet, the right-handed neutrino.

6.6 Calculate the β-function coefficients in the MSSM (eq. (6.3)) as a function of the number of Higgs doublets. Rewrite the one-loop solutions and the one-loop predictions for t, M_U and $\alpha_3(M_Z)$ as a function of the number of Higgs doublets as well. Examine the variation of the predictions as you decrease/increase that number. The sharp change in the predictions is because the Higgs doublets do not form a complete GUT representations. Add the down-type color triplet contribution (so that the extra Higgs doublet are embedded in $5 + \bar{5}$ of $SU(5)$, e.g. the messenger fields in Sect. 10.3) to the β-functions and repeat the exercise. The triplet completes the doublet GUT representation.

6.7 Solve the gauge part of the Yukawa unification equation (6.8) and show that in the limit of small Yukawa couplings $y \ll g$

$$\left(\frac{y_d}{y_l}\right)\bigg|_{M_Z} = \left(\frac{\alpha_3(M_Z)}{\alpha_G}\right)^{\frac{8}{9}} \left(\frac{\alpha_1(M_Z)}{\alpha_G}\right)^{\frac{10}{90}}. \tag{6.10}$$

Compare the predictions for m_d/m_e and m_s/m_μ (are they distinguished?) to the data. This is essentially the fermion mass problem in GUT's.

6.8 Introduce group theory (QCD and $SU(2)$) as well as generation indices to the proton decay operator $QQQL$ and show that one Q must be of the second generation or otherwise the group theory forces the operators to vanish once antisymmetric indices are properly summed. (Why not third generation?) Write the corresponding one-loop (box) proton decay diagram with a gaugino, a colored Higgsino, and sfermions circulating in the loop and derive the proportionality relation given in the text. (Note that the two colored Higgsinos form a heavy Dirac fermion.)

7. The Heavy Top and Radiative Symmetry Breaking

Now that we are familiar with the notion and practice of renormalization group evolution, we will continue and renormalize the MSSM in order to understand the possible ultraviolet origins of its infrared structure. (For example, see Ref. [92].) We begin with an intriguing feature of the equations, the quasi-fixed point. We then discuss the renormalization of the spectrum parameters. Though we will use the quasi-fixed point regime for demonstration, this is done for simplicity only and the radiative-symmetry-breaking results derived below apply in general.

7.1 The Quasi-Fixed Point

Aside from gauge couplings, the top mass, and hence the top Yukawa (baring in mind the possibility of $\pm \mathcal{O}(5-10\%)$ finite corrections from sparticle loops), is also measured quite precisely, $y_t(m_t) \simeq 0.95/\sin\beta$. Again, one can ask whether such a large weak-scale coupling remains perturbative (when properly renormalized) up to the unification scale. The answer is positive for the following reason: Upon examining the one-loop renormalization group equation for y_t,

$$\frac{dy_t}{d\ln\Lambda} \simeq \frac{y_t}{16\pi^2}\left\{-\frac{16}{3}g_3^2 + 6y_t^2\right\}, \tag{7.1}$$

where $g_3 \equiv g_s$ is the QCD gauge coupling and we neglected all other couplings, one finds a semi fixed-point behavior [93]. For $y_{t_\text{fixed}}^2 = (16/18)g_3^2 \simeq (1.15)^2$ the right hand side equals zero (where we used, as an approximation, the weak-scale value of g_3) and y_t freezes at this value and is not renormalized any further.

For large enough values of $y_t(M_U)$, for example, y_t decreases with energy according to eq. (7.1) until it reaches y_{t_fixed}. This is the top-Yukawa quasi-fixed-point value, i.e., convergence from above. The quasi-fixed-point behavior is illustrated in Fig. 7.1. (The value of y_{t_fixed} diminishes if there are any other large Yukawa couplings, for example a large y_b, right-handed neutrino couplings, R_P-violating couplings, singlet couplings in the NMSSM, etc., which modify eq. (7.1): $\sum c_i y_{i_\text{fixed}}^2 = (16/18)g_3^2$ for some coefficients c_i.) On the other hand, weak-scale values $y_t > y_{t_\text{fixed}}$ imply $y_t \gg 1$ at intermediate energies below the unification scale. This then gives a lower bound on

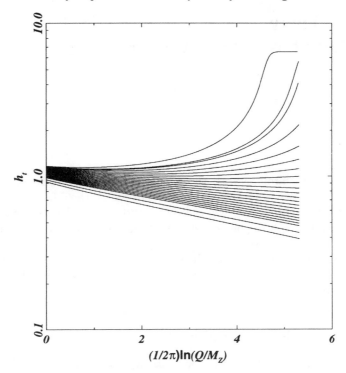

Fig. 7.1. Convergence of the the top Yukawa coupling (denoted here h_t) to its quasi-fixed point value, given a large range of initial values at the unification scale, is illustrated (at two-loop order) as a function of the evolution time t, $t = 0$ at M_Z. For very large initial values perturbative validity does not hold and the curves are shown for illustration only. (The flat behavior is a two-loop effect.) Note the logarithmic scale.

$\tan \beta \gtrsim 1.8$, referred to following eq. (5.1). (An upper bound $\tan \beta \lesssim 60$ is derived by applying similar consideration to the bottom-Yukawa coupling.) Renormalization group study, however, shows that the fixed point is not reached for any (arbitrarily small) initial value of $y_t(M_U)$, and in fact, even when it is reached it is not an exact value. Rather, a small $\mathcal{O}(1\%)$ sensitivity for the boundary conditions remains. Hence, it is a quasi-fixed point[1] [93] (which serves more as an upper bound to insure consistency of perturbation theory up to the unification scale). Nevertheless, its existence can beautifully explain $y_t(m_t) \sim 1$ for a large range of initial values at the unification scale. It is this behavior that plays a crucial role in the successful prediction of $b - \tau$ unification discussed in Sect. 6.2.1.

---

[1] Note that the equations also contain a true fixed point [94] which, however, is not reached given the values of the physical parameters.

Current bounds on the Higgs mass, however, limit $\tan \beta$ from below $\tan \beta \gtrsim 2$ (see the next chapter) so that the quasi-fixed point scenario may not be fully realized, at least not in its simplest version. Nevertheless, whether or not the top-Yukawa saturates its upper bound, given the heavy top mass it is a large coupling. Its renormalization curve is attracted[2] towards the quasi-fixed point and $y_t(\Lambda) \simeq \mathcal{O}(1)$ even though it may not actually reach the fixed point. (See Fig. 7.1.) This is an essential and necessary ingredient in the renormalization of the SSB parameters from some (at this point, arbitrary) boundary conditions, for example, at the unification scale. (It is implicitly assumed here that these parameters appear in the effective theory and are "hard" already at high-energies, but we will return to this point in Chap. 10.)

Next, we focus (only) on the renormalization of the terms relevant for electroweak symmetry breaking; the radiative symmetry breaking (RSB) mechanism. A large y_t coupling was postulated more than a decade ago as a mean to achieve RSB [95, 96] and in that sense is a prediction of the MSSM framework which is successfully confirmed by the experimentally measured heavy top mass $m_t \simeq \nu$.

7.2 Radiative Symmetry Breaking

In order to reproduce the SM Lagrangian properly, a negative mass squared in the Higgs potential is required. (See also Chap. 12.) Indeed, the $m_{H_2}^2$ parameter is differentiated from all other squared mass parameters once we include the Yukawa interactions. Consider the coupled renormalization group equations, including, for simplicity, only gauge and top-Yukawa effects. (More generally, the b-quark, τ-lepton, right-handed neutrino, singlet and R_P-violating couplings may not be negligible.) Then, the one-loop evolution of $m_{H_2}^2$ (and of the coupled parameters $m_{U_3}^2$ and $m_{Q_3}^2$) with respect to the logarithm of the momentum-scale is given by

$$\frac{dm_{H_2}^2}{d \ln \Lambda} = \frac{1}{8\pi^2}(3y_t^2 \Sigma_{m^2} - 3g_2^2 M_2^2 - g_1^2 M_1^2), \tag{7.2}$$

and

$$\frac{dm_{U_3}^2}{d \ln \Lambda} = \frac{1}{8\pi^2}(2y_t^2 \Sigma_{m^2} - \frac{16}{3}g_3^2 M_3^2 - \frac{16}{9}g_1^2 M_1^2), \tag{7.3}$$

$$\frac{dm_{Q_3}^2}{d \ln \Lambda} = \frac{1}{8\pi^2}(y_t^2 \Sigma_{m^2} - \frac{16}{3}g_3^2 M_3^2 - 3g_2^2 M_2^2 - \frac{1}{9}g_1^2 M_1^2), \tag{7.4}$$

where $\Sigma_{m^2} = [m_{H_2}^2 + m_{Q_3}^2 + m_{U_3}^2 + A_t^2]$, and we denote the SM $SU(3)$, $SU(2)$ and (the GUT normalized) $U(1)$ gaugino masses by $M_{3,2,1}$, as before.

---

[2] It can be shown that more generally $y_t(m_t) \simeq \mathcal{O}(1)$ for $y_t(M_U) \gtrsim 0.5$. See Fig. 7.1.

The one-loop gaugino mass renormalization obeys

$$\frac{d}{d\ln \Lambda}\left(\frac{M_i^2}{g_i^2}\right) = 0, \tag{7.5}$$

and its solution simply reads $M_i(M_i) = (\alpha_i(M_i)/\alpha_i(\Lambda_{\mathrm{UV}}))M_i(\Lambda_{\mathrm{UV}})$ where a typical choice is $\Lambda_{\mathrm{UV}} = M_U$. Note that in unified theories the gaugino mass boundary conditions are given universally by the mass of the single gaugino of the GUT group so that $M_3 : M_2 : M_1 \simeq 3 : 1 : 1/2$ at the weak scale (where the numerical ratios are the ratios $\alpha_i(M_Z)/\alpha_U$). This is gaugino mass unification. (It also holds, but for different reasons, in many string models.)

Given the heavy t-quark, one has $y_t \sim 1 \sim g_3$. (In fact, for near quasi-fixed point values typically $y_t > g_3$ at high energies.) While QCD loops still dominate the evolution of the stop masses squared $m_{Q_3}^2$ and $m_{U_3}^2$, Yukawa loops dominate the evolution of $m_{H_2}^2$. On the one hand, the stop squared masses and Σ_{m^2} increase with the decreasing scale. On the other hand, the more they increase the more the Higgs squared mass decreases with scale and, given the integration or evolution time $2\pi t \sim 30$, it is rendered negative at or before the weak scale. (This mechanism hold more generally for different values of t. As t decreases the ratio $|m_{Q,U}^2/m_{H_2}^2|$ must increase.) The $m_3^2 H_1 H_2$ Higgs doublet mixing term ensures that both Higgs doublets have non-vanishing expectation values.

This is a simplistic description of the mechanism of radiative electroweak symmetry breaking. In fact, the sizeable y_t typically renders the Higgs squared mass too negative and some (fine?) tuning (typically of μ) is required in order to extract correctly the precisely known electroweak scale. The degree of acceptable tuning can be argued to bound the mass parameters from above, e.g. $m_{\tilde{t}} \lesssim (4\pi/y_t)M_Z$, which is a rephrasing of the generic bound eq. (5.14). (The Higgs potential and its minimization are discussed in the next chapter.) An example of the renormalization group evolution of the SSB (and μ) parameter is illustrated in Fig. 7.2, taken from C. Kolda. For illustration, universal (see Chap. 10) boundary conditions are assumed at the unification scale M_U.

In the quasi-fixed point scenario, it is possible to solve analytically for the low energy values of the soft scalar masses in terms of the high scale boundary conditions. For illustration, we conclude our discussion with those solutions. We include, for completeness, also the solutions for sfermions and Higgs SSB masses which are not affected by the large y_t (for example, see Carena et $al.$ in Ref. [87]):

$$m_{H_2}^2 \simeq m_{H_2}^2(M_U) + 0.52M_{1/2}^2 - 3\Delta m^2$$
$$m_{H_1}^2 \simeq m_{H_1}^2(M_U) + 0.52M_{1/2}^2$$
$$m_{Q_i}^2 \simeq m_{Q_i}^2(M_U) + 7.2M_{1/2}^2 - \delta_i \Delta m^2$$
$$m_{U_i}^2 \simeq m_{U_i}^2(M_U) + 6.7M_{1/2}^2 - \delta_i 2\Delta m^2 \tag{7.6}$$

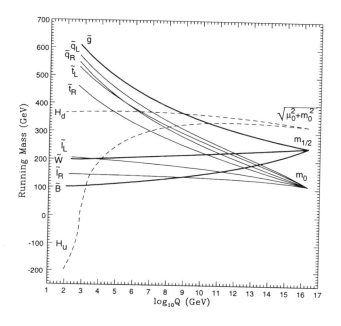

Fig. 7.2. The renormalization group evolution of SSB masses and μ for a representative case with universal boundary conditions at the unification scale M_U. $\tilde{t}_L = \widetilde{Q}_3$ and $\tilde{t}_R = \widetilde{U}_3$, and $\tilde{q}_{L\,(R)}$ is a left- (right-) handed squark of the first two generations.

$$m^2_{D_i} \simeq m^2_{D_i}(M_U) + 6.7M^2_{1/2}$$
$$m^2_{L_i} \simeq m^2_{L_i}(M_U) + 0.52M^2_{1/2}$$
$$m^2_{E_i} \simeq m^2_{E_i}(M_U) + 0.15M^2_{1/2} \ ,$$

where

$$\Delta m^2 \simeq \frac{1}{6}\left[m^2_{H_2}(M_U) + m^2_{Q_3}(M_U) + m^2_{U_3}(M_U)\right] r$$
$$+ M^2_{1/2}\left(\frac{7}{3}r - r^2\right) + \frac{1}{3}A_0\left(\frac{1}{2}A_0 - 2.3M_{1/2}\right)r\,(1-r) \ , \quad (7.7)$$

and, for simplicity, and as is customary, we have assumed a common gaugino mass $M_{1/2}$ and trilinear scalar coupling A_0 at the high scale, which is conveniently identified here with the unification scale M_U. (Numerical coefficients will change otherwise) We will not discuss in detail the renormalization of the A-parameters, which are assumed here to be proportional to the Yukawa couplings. However, we note that their renormalization also exhibits fixed points. The subscript i is a generational index; $\delta_1 = \delta_2 = 0$ and $\delta_3 = 1$.

Finally, the parameter $r = [y_t/y_{t_{\text{fixed}}}]^2 \leq 1$ is a measure of the proximity of the top Yukawa coupling to its quasi-fixed point value at the weak scale.

It is interesting to observe that for vanishing gaugino masses the system (7.2)–(7.4) has a zero fixed point (*i.e.*, it is insensitive to the value of m_0^2) as long as $[m_{H_2}^2(M_U), m_{U_3}^2(M_U), m_{Q_3}^2(M_U)] = m_0^2\ [3, 2, 1]$. (See exercise.) Such fixed points, in the presence of Yukawa quasi-fixed points, are a more general phenomenon with important consequences for the upper bound on m_0 [97] and hence, "(fine-)tuning". This also relates to the concept of focus points [98] which may be similarly realized (*i*) for particular boundary-condition patterns, (*ii*) and for particular values of y_t (which experimentally coincide with intermediate values of $\tan\beta \gtrsim 10$), (*iii*) and if the ultraviolet scale is (or near) the unification scale. If the "focusing" is realized then $m_{H_2}^2(M_Z) \simeq -M_Z^2$ regardless of the exact magnitude of its boundary condition at the unification scale. The focus point behavior also undermines (fine-)tuning considerations to some degree. Both, the fixed and the focus points, hold at one loop and their application is constrained by the two-loop corrections.

The general (two-loop) renormalization group equations are given in Ref. [99]. Other related papers include Ref. [100].

Exercises

7.1 Confirm the one-loop gaugino-mass unification relation $M_3 : M_2 : M_1 \simeq 3 : 1 : 1/2$ at the weak scale.

7.2 Omitting all terms aside from QCD in the stop renormalization group equation, and using the renormalization group equations for the gauge couplings and for the gaugino mass, show that if $m_Q(\Lambda_{\text{UV}}) = 0$ then $m_Q^2(m_Q) \simeq 7M_3^2(\Lambda_{\text{UV}}) \simeq M_3^2(M_3)$, where m_Q correspond to the stop doublet mass and M_3 to the gluino mass. Compare to the analytic solution with $\delta_i = 0$. Include all gauge terms and assume gaugino mass unification, can you derive the difference between the renormalized m_Q^2 and m_U^2 ?

7.3 Write the one-loop renormalization group equations (7.2)–(7.4) in a matrix form. This defines a three-dimensional space. Show that the vector $[m_{H_2}^2(\Lambda_{\text{UV}}), m_{U_3}^2(\Lambda_{\text{UV}}), m_{Q_3}^2(\Lambda_{\text{UV}})] = m_0^2\ [3, 2, 1]$ is an eigenvector of the coefficient matrix in the limit of vanishing gaugino masses (and $A_0 = 0$). Find its eigenvalue and solve the matrix equation along this direction for a fixed y_t. Compare to eq. (7.6). Study the behavior of the solution and show that it has a zero stop-Higgs fixed point in this case. In practice, it implies that the gaugino mass, rather than the sfermion boundary conditions, dictate the low-energy value of the vector. Find all eigenvalues and eigenvectors and write the general solution in this limit.

7.4 Generalize eqs. (7.2)–(7.4) to the case $y_t \simeq y_b \simeq \mathcal{O}(1)$, as is appropriate in the large $\tan\beta$ regime. (Note: $t \to b$, $Q \to Q$, $U \to D$, $H_2 \to H_1$.) Next, generalize eq. (7.6) accordingly. (For simplicity, denote $\Delta m^2 \to \Delta m_t^2$, Δm_b^2.) Finally, repeat Ex. 7.3 for this case (assume $y_t = y_b$). This exercise can be further generalized for the "$SO(10)$ unification" case, $y_{\nu_\tau} = y_t = y_b = y_\tau$.

7.5 Consider the scale-dependent behavior of $m_{H_2}^2(\Lambda)$ in the case $m_{H_2}^2(M_U) < 0$. (M_U serves as the ultraviolet boundary.) Show that, in general, $m_{H_2}^2(\Lambda)$ is bounded from below.

8. The Higgs Potential and the Light Higgs Boson

In the previous chapter it was demonstrated that a negative mass squared in the Higgs potential is generated radiatively for a large range of boundary conditions. We are now in position to write and minimize the Higgs potential and examine the mass eigenvalues and eigenstates and their characteristics. We will do so in this chapter.

8.1 Minimization of the Higgs Potential

In principle, it is far from clear that the Higgs bosons rather than some sfermion receive vev's. Aside from the sneutrino (whose vev only breaks lepton number, leading to the generation of neutrino masses) all other sfermions cannot have non-vanishing expectation values or otherwise QED and/or QCD would be spontaneously broken. Furthermore, there could be some direction in this many field space in which the complete scalar potential (which involves all Higgs and sfermion fields) is not bounded from below. We reserve the discussion of these issues to Chap. 12 and only state here that they lead to constraints on the parameter space, for example, on the ratios of the SSB parameters $|A_f/m_{\tilde{f}}|^2 \lesssim 3 - 6$. For the purpose of this chapter, let us simply assume that any such constraints are satisfied and let us focus on the Higgs potential.

The Higgs part of the MSSM (weak-scale) scalar potential reads, assuming for simplicity that all the parameters are real,

$$V(H_1, H_2) = (m_{H_1}^2 + \mu^2)|H_1|^2 + (m_{H_2}^2 + \mu^2)|H_2|^2$$
$$-m_3^2(H_1 H_2 + h.c.) + \frac{\lambda^{\text{MSSM}}}{2}(|H_2|^2 - |H_1|^2)^2 + \Delta V, \qquad (8.1)$$

where $m_{H_1}^2$, $m_{H_2}^2$, and m_3^2 (μ) are the soft (supersymmetric) mass parameters renormalized down to the weak scale (*i.e.*, eq. (8.1) is the one-loop improved tree-level potential), $m_3^2 > 0$, $\lambda^{\text{MSSM}} = (g_2^2 + g'^2)/4$ is given by the hypercharge and $SU(2)$ D-terms, and we suppress $SU(2)$ indices. (Note that F-terms do not contribute to the pure Higgs quartic potential in the MSSM. They do contribute in the NMSSM.) The one-loop correction $\Delta V = \Delta V^{\text{one-loop}}$ (which, in fact, is a threshold correction to the one-loop improved tree-level potential) can be absorbed to a good approximation in redefinitions of the tree-level parameters.

A broken $SU(2) \times U(1)$ (along with the constraint $m_{H_1}^2 + m_{H_2}^2 + 2\mu^2 \geq 2|m_3^2|$ from vacuum stability, $i.e.$, the requirement of a bounded potential along the flat $\tan \beta = 1$ direction) requires

$$(m_{H_1}^2 + \mu^2)(m_{H_2}^2 + \mu^2) \leq |m_3^2|^2. \tag{8.2}$$

Eq. (8.2) is automatically satisfied for $m_{H_2}^2 + \mu^2 < 0$ (given $m_{H_1}^2 > 0$), which was the situation discussed in Chap. 7. The minimization conditions then give

$$\mu^2 = \frac{m_{H_1}^2 - m_{H_2}^2 \tan^2 \beta}{\tan^2 \beta - 1} - \frac{1}{2} M_Z^2, \tag{8.3}$$

$$m_3^2 = -\frac{1}{2} \sin 2\beta \left[m_{H_1}^2 + m_{H_2}^2 + 2\mu^2 \right]. \tag{8.4}$$

By writing eq. (8.3) we subscribed to the convenient notion that μ is determined by the precisely known value $M_Z = 91.19$ GeV. This is a mere convenience. Renormalization cannot mix the supersymmetric μ parameter (which is protected by non-renormalization theorems which apply to the superpotential, $d\mu/d \ln \Lambda \propto \mu$) and the SSB parameters, and hence the independent μ can be treated as a purely low-energy parameter. Nevertheless it highlights the μ-problem, why is a supersymmetric mass parameter exactly of the order of magnitude of the SSB parameters (rather than $\Lambda_{\rm UV}$, for example) [101]. We touched upon this point in the context of GUT's and doublet-triple splitting, but it is a much more general puzzle whose solution must encode some information on the ultraviolet theory which explains this relation. (Several answers were proposed in the literature, including Ref. [101, 102, 103, 104] and various variants of the NMSSM.)

The above form of eq. (8.3) also highlights the fine-tuning issue whose rough measure is the ratio $|\mu/M_Z|$. Typically $|m_{H_2}^2|$ is a relatively large parameter controlled by the stop renormalization, which itself is controlled by QCD and gluino loops. One often finds that a phenomenologically acceptable value of μ is $|\mu(M_Z)| \simeq |M_3(M_Z)|$ and that M_Z is then determined by a cancellation between two $\mathcal{O}(\text{TeV})$ parameters, e.g. $(1/2)M_Z^2 \simeq -(m_{H_2}^2 + \mu^2)$ in the large $\tan \beta$ limit. Clearly, this is a product of our practical decision to fix M_Z rather than extract it. All it tells us is that M_Z (or ν) is a special rather than arbitrary value. The true tuning problem is instead in the relation $|\mu| \simeq |M_3|$ which is difficult to understand. Fine-tuning is difficult to quantify, and each of its definitions in the literature has its own merits and conceptual difficulties. Caution is in place when applying such esthetic notions to actual calculations, an application which we will avoid.

8.2 The Higgs Spectrum and Its Symmetries

Using the minimization equations, the pseudo-scalar mass-squared matrix (5.21) (the corresponding CP-even and charged Higgs matrices receive also contribution from the D-terms, or equivalently, from the quartic terms) is now

$$M_{PS}^2 = m_3^2 \begin{pmatrix} \tan\beta & -1 \\ -1 & 1/\tan\beta \end{pmatrix}. \tag{8.5}$$

Its determinant vanishes due to the massless Goldstone boson. It has a positive mass-squared eigenvalue $m_{A^0}^2 = \mathrm{Tr} M_{PS}^2 = m_3^2/((1/2)\sin 2\beta) = m_1^2 + m_2^2$, where as before $m_i^2 \equiv m_{H_i}^2 + \mu^2$ for $i = 1, 2$. Electroweak symmetry breaking is then confirmed. The angle β is now seen to be the rotation angle between the current and mass eigenstates.

The CP-even Higgs tree-level mass matrix reads

$$M_{H^0}^2 = m_{A^0}^2 \begin{pmatrix} s_\beta^2 & -s_\beta c_\beta \\ -s_\beta c_\beta & c_\beta^2 \end{pmatrix} + M_Z^2 \begin{pmatrix} c_\beta^2 & -s_\beta c_\beta \\ -s_\beta c_\beta & s_\beta^2 \end{pmatrix}, \tag{8.6}$$

with (tree-level) eigenvalues

$$m_{h^0, H^0}^{2\,T} = \frac{1}{2}\left[m_{A^0}^2 + M_Z^2 \mp \sqrt{(m_{A^0}^2 + M_Z^2)^2 - 4 m_{A^0}^2 M_Z^2 \cos^2 2\beta} \right]. \tag{8.7}$$

Note that at this level there is a sum rule for the neutral Higgs eigenvalues: $m_{H^0}^2 + m_{h^0}^2 = m_{A^0}^2 + M_Z^2$.

There are two particularly interesting limits to eq. (8.7). In the limit $\tan\beta \to 1$ one has $|\mu| \to \infty$ and the $SU(2) \times U(1)$ breaking is driven by the m_3^2 term. In practice, one avoids the divergent limit by taking $\tan\beta \gtrsim 1.1$, as is also required from perturbativity of the top-Yukawa coupling and by the experimental lower bound on the Higgs boson mass (discussed in the next section). For $\tan\beta \to \infty$ one has $m_3 \to 0$ so that the symmetry breaking is driven by $m_{H_2}^2 < 0$.

The $\tan\beta \to 1$ case corresponds to an approximate $SU(2)_{L+R}$ custodial symmetry of the vacuum: Turning off hypercharge and flavor mixing, and if $y_t = y_b = y$, then one can rewrite the t and b Yukawa terms in a $SU(2)_L \times SU(2)_R$ invariant form [105],

$$y \begin{pmatrix} t_L \\ b_L \end{pmatrix}_a \epsilon_{ab} \begin{pmatrix} H_1^0 & H_2^+ \\ H_1^- & H_2^0 \end{pmatrix}_{bc} \begin{pmatrix} -b_L^c \\ t_L^c \end{pmatrix}_c \tag{8.8}$$

where in the SM $H_2 = i\sigma_2 H_1^*$. For $v_1 = v_2$ (as in the SM or in the $\tan\beta \to 1$ limit) the symmetry is spontaneously broken $SU(2)_L \times SU(2)_R \to SU(2)_{L+R}$. However, $y_t \neq y_b$ and the different hypercharges of $U_3 = t_L^c$ and $D_3 = b_L^c$ explicitly break the left-right symmetry, and therefore the residual custodial symmetry.

In the MSSM, on the other hand, H_1 is distinct from H_2 and if $\nu_1 \neq \nu_2$ (where $\nu_i = \langle H_i^0 \rangle$) $SU(2)_L \times SU(2)_R \to U(1)_{T_{3L}+T_{3R}}$. Therefore, the $SU(2)_{L+R}$ symmetry is preserved if $\beta = \frac{\pi}{4}$ ($\nu_1 = \nu_2$) and is maximally broken if $\beta = \frac{\pi}{2}$ ($\nu_1 \ll \nu_2$). (This is the same approximate custodial symmetry which was mentioned above in the context of the smallness of quantum corrections to electroweak observables and couplings, but as manifested in the Higgs sector.) The symmetry is broken at the loop level so that one expects in any case $\tan\beta$ above unity. As a result of the symmetry,

$$M_{H^0}^2 \approx \mu^2 \times \begin{pmatrix} 1 & -1 \\ -1 & 1 \end{pmatrix}, \tag{8.9}$$

and it has a massless tree-level eigenvalue, $m_{h^0}^T \approx 0$. This is, of course, a well known result of the tree-level formula when taking $\beta \simeq \frac{\pi}{4}$. The mass is then determined by the loop corrections which are well known (to two-loop) $m_{h^0}^2 \approx \Delta_{h^0}^2 \propto h_t^2 m_t^2$ (see the next section). The heavier CP-even Higgs boson mass eigenvalue equals approximately $\sqrt{2}|\mu|$. (The loop corrections are less relevant here as typically $m_{H^0}^2 \gg \Delta_{H^0}^2$). The custodial symmetry (or the large μ parameter) dictates in this case a degeneracy $m_{A^0} \approx m_{H^0} \approx m_{H^+} \approx \sqrt{2}|\mu|$. (The tree-level corrections to that relation are $\mathcal{O}(M_{W,Z}/m_{A^0})^2$.) That is, at a scale $\Lambda \approx \sqrt{2}|\mu|$ the heavy Higgs doublet H is decoupled, and the effective field theory below that scale has only one SM-like ($\nu_{h^0} = \nu$) Higgs doublet, h ($= H$ of Chap. 1) which contains a light physical state. This is a special case of the MSSM in which all other Higgs bosons (and possibly sparticles) decouple. (The decoupling limit typically holds for $m_{A^0} \gtrsim 300$ GeV, and is realized more generally. See exercise.)

The Higgs sector in the large $\tan\beta$ case exhibits an approximate $O_4 \times O_4$ symmetry [106]. For $m_3 \to 0$ (which is the situation in case (2)) there is no mixing between H_1 and H_2 and the Higgs sector respects the $O_4 \times O_4$ symmetry (up to gauge-coupling corrections), i.e., invariance under independent rotations of each doublet. The symmetry is broken to $O_3 \times O_3$ for $\nu_1 \neq \nu_2 \neq 0$ and the six Goldstone bosons are the three SM Goldstone bosons, A^0, and $H^\pm$. The symmetry is explicitly broken for $g_2 \neq 0$ (so that $m_{H^+} = M_W$) and is not exact even when neglecting gauge couplings (i.e., $m_3 \neq 0$). Thus, A^0 and $H^\pm$ are massive pseudo-Goldstone bosons, $m_{H^+}^2 - M_W^2 \approx m_{A^0}^2 = C \times m_3^2$. However, $C = -2/\sin 2\beta$ and it can be large, which is a manifestation of the fact that $O_4 \times O_4 \to O_4 \times O_3$ for $\nu_1 = 0$. (The limit $m_3 \to 0$ corresponds also to a $U(1)$ Peccei-Quinn symmetry under which the combination $H_1 H_2$ is charged.) In the case $\beta \to \frac{\pi}{2}$ one has $m_{h^0}^T \approx M_Z$ (assuming $m_{A^0} \geq M_Z$). When adding the loop corrections $m_{h^0} \lesssim \sqrt{2}M_Z \approx 130$ GeV. (See the next section.)

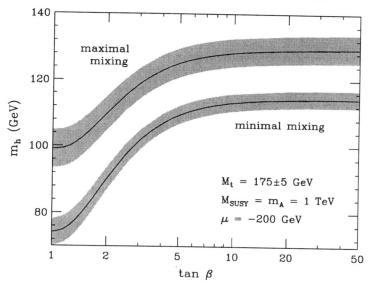

Fig. 8.1. The radiatively corrected light CP-even Higgs mass is plotted as a function of $\tan\beta$, for the maximal squark left-right mixing (upper band) and minimal squark mixing cases. The impact of the top-quark mass is exhibited by the shaded bands; the central value corresponds to $m_t = 175$ GeV, while the upper (lower) edge of the bands correspond to increasing (decreasing) m_t by 5 GeV. M_{SUSY} is a typical superpartner mass and μ is the Higgsino mass parameter. Taken from Ref. [114].

8.3 The Light Higgs Boson

Before concluding the discussion of the Higgs sector, let us examine the lightness of the SM-like Higgs boson from a different perspective, as well as the one-loop corrections to its mass. Including one-loop corrections, the general upper bound is derived

$$m_{h^0}^2 \leq M_Z^2 \cos^2 2\beta + \frac{3\alpha m_t^4}{4\pi s^2 (1-s^2) M_Z^2} \left\{ \ln\left(\frac{m_{\tilde{t}_1}^2 \, m_{\tilde{t}_2}^2}{m_t^4}\right) + \Delta_{\theta_t} \right\} \qquad (8.10)$$

where

$$\Delta_{\theta_t} = \left(m_{\tilde{t}_1}^2 - m_{\tilde{t}_2}^2\right) \frac{\sin^2 2\theta_t}{2m_t^2} \ln\left(\frac{m_{\tilde{t}_1}^2}{m_{\tilde{t}_2}^2}\right)$$
$$+ \left(m_{\tilde{t}_1}^2 - m_{\tilde{t}_2}^2\right)^2 \left(\frac{\sin^2 2\theta_t}{4m_t^2}\right)^2 \left[2 - \frac{m_{\tilde{t}_1}^2 + m_{\tilde{t}_2}^2}{m_{\tilde{t}_1}^2 - m_{\tilde{t}_2}^2} \ln\left(\frac{m_{\tilde{t}_1}^2}{m_{\tilde{t}_2}^2}\right)\right], \qquad (8.11)$$

and where $m_{\tilde{t}_i}^2$ are the eigenvalues of the stop $\tilde{t}$ mass-squared matrix, θ_t is the left-right stop mixing angle, and we have neglected other loop contributions. The tree-level mass squared, $m_{h^0}^{T\,2}$, and the loop correction, $\Delta_{h^0}^2$, are bounded by the first and second terms on the right-hand side of eq. (8.10), respectively. In the absence of mixing, $\Delta_{\theta_t} = 0$. For $\tan \beta \to 1$ one obtains for the tree-level mass $m_{h^0}^{2T} \to 0$, and thus $m_{h^0}^2 \approx \Delta_{h^0}^2$.

Clearly, and as we observed before, the tree-level mass vanishes as $\tan \beta \to 1$ ($\cos 2\beta \to 0$). In this limit, the D-term (expectation value) vanishes as well as the value of the tree-level potential which is now quadratic in the fields. It corresponds to a flat direction of the potential and the massless real-scalar h^0 is its ground state. Now that we have identified the flat direction it is clear that the upper bound must be proportional to $\cos 2\beta$ so that h^0, which parameterizes this direction, is massless once the flat direction is realized. The proportionality to M_Z is only the manifestation that the quartic couplings are the gauge couplings. Hence, the lightness of the Higgs boson is a model-independent statement.

The flat direction is always lifted by quantum corrections, the most important of which is given in eq. (8.10). These corrections may be viewed as effective quartic couplings that have to be introduced to the effective theory once the stops $\tilde{t}_i$, for example, are integrated out of the theory at a few hundred GeV or higher scale. These couplings are proportional to the large Yukawa couplings (for example from integrating out loops induced by $(y_t \tilde{t} H_i)^2$ quartic F-terms in the scalar potential, as in Fig. 2.3). Note that even though one finds in many cases $\mathcal{O}(100\%)$ corrections to the light Higgs mass (and hence $m_{h^0} \leq M_Z \to m_{h^0} \lesssim \sqrt{2} M_Z$) this does not signal the breakdown of perturbation theory. It is only that the tree-level mass (approximately) vanishes. Indeed, two-loop corrections are much smaller (and shift m_{h^0} by typically only a few GeV) and are often negative.

The above upper bound is modified if and only if the Higgs potential contains terms (aside from the loop corrections) that lift the flat direction, for example, this is the case in the NMSSM (see, for example, Ref. [107, 108], and references therein) or if the gauge structure is extended by an Abelian factor [109] SM $\to$ SM$\times U(1)'$ (an extension that could still be consistent with gauge coupling unification [110] though this is not straightforward). However, as long as one requires all coupling to stay perturbative in the ultraviolet then the additional contributions to the Higgs mass are still modest leading to $m_{h^0} \lesssim 150 - 200$ GeV (including loop corrections), where the upper bound of $190 - 200$ GeV was shown [111] to be saturated only in somewhat contrived constructions (which still preserve unification). In certain supersymmetric SM $\to$ SM$\times U(1)'$ frameworks (which contain also SM singlets) SM-like Higgs boson as heavy as 180 GeV (including loop corrections) was found [112]. This corresponds to roughly $\mathcal{O}(100\%) \simeq M_Z$ corrections due to a perturbative singlet and $\mathcal{O}(100\%)$ corrections from a weakly coupled Abelian factor, so that (adding in quadrature) $m_{h^0}^2 \lesssim 4 M_Z^2$. The only known caveat in the argument for model-dependent lightness of the Higgs boson is the case of low-energy supersymmetry breaking where new tree-level terms may appear in the Higgs potential [78]. (See Sect. 13.2.)

The existence of a model-independent light Higgs boson is therefore a prediction of the framework (with the above caveat). It is encouraging to note that it seems to be consistent with current data. The W mass measurement and other electroweak observable strongly indicate that the SM-like Higgs is light $m_{h^0} \lesssim 200\text{-}300$ GeV where the best fitted values are near 100 GeV [13, 84]. (Some caveats, however, remain [14, 15], as discussed in Chap. 1.) Searches at the LEP experiments bound the SM-like Higgs mass from below $m_{h^0} \gtrsim 115$ GeV [113]. As noted earlier (see Sect. 2.3), both theory and experiment indicate a (SM-like) Higgs mass range

$$g_{\text{Weak}}\nu \lesssim m_{h^0} \lesssim \nu \tag{8.12}$$

(where $g_{\text{Weak}}\nu \equiv M_Z$). This is an encouraging hint and an important consistency check. Depending on final luminosity, the model-independent light Higgs may be probed already in the current Tevatron run, but it may be that its discovery (or exclusion) must wait until the LHC is operative[1].

The predicted mass range for the light Higgs boson in the MSSM is illustrated in Fig. 8.1 taken from Ref. [114]. Further Discussion of the Higgs sector and references can also be found, for example, in Refs. [115, 116].

Exercises

8.1 Derive the minimization conditions of the Higgs potential given above by first organizing the minimization equations as

$$m_1^2 = m_3^2 \frac{\nu_2}{\nu_1} + \frac{1}{4} \left(g_2^2 + g'^2 \right) \left(\nu_2^2 - \nu_1^2 \right) \tag{8.13}$$

$$m_2^2 = m_3^2 \frac{\nu_1}{\nu_2} + \frac{1}{4} \left(g_2^2 + g'^2 \right) \left(\nu_2^2 - \nu_1^2 \right), \tag{8.14}$$

where $m_i^2 \equiv m_{H_i}^2 + \mu^2$ for $i = 1, 2$.

8.2 Derive the tree-level charged Higgs mass-squared matrix $M_{H^\pm}^2 = M_{PS}^2(1 + (1/2)M_W^2 \sin 2\beta)$, show that it has a massless eigenvalue (the Goldstone boson) and derive the sum rule $m_{H^\pm}^2 = m_{A^0}^2 + M_W^2$.

---

[1] Detection of a Higgs boson in this mass range may be difficult in the LHC environment as it relies on the suppressed radiative diphoton decay channel $h^0 \to \gamma\gamma$.

8.3 Show that the rotation angle of the CP-even Higgs states is given by

$$\sin 2\alpha = -\frac{m_{A^0}^2 + M_Z^2}{m_{H^0}^2 - m_{A^0}^2} \sin 2\beta. \qquad (8.15)$$

One can define a decoupling limit $\alpha \to \beta - (\pi/2)$, which is reached, in practice, for $m_{A^0} \gtrsim 300$ GeV. Show that in this limit the heavy Higgs eigenstates form a doublet H that does not participate in electroweak symmetry breaking. The other (SM-like) Higgs doublet is roughly given in this limit (up to $\mathcal{O}(M_Z^2/M_H^2)$ corrections and a phase) by $h \simeq H_1 \cos\beta + H_2 \sin\beta$.

8.4 Derive the tree-level bound $m_{h^0} \leq M_Z |\cos 2\beta|$ from eq. (8.7).

8.5 Use the experimental lower bound on the mass of a SM-like Higgs boson and the Higgs mass formulae to derive a lower bound on $\tan\beta$. Repeat for various values of the parameters entering the loop correction. Show that the 2000 experimental lower bound [113] quoted above implies $\tan\beta \gtrsim 2$. For caveats, see Ref. [78].

8.6 In a common version of the NMSSM, $W = \mu H_1 H_2$ is replaced with $W = \lambda S H_1 H_2 + (\kappa/3) S^3$, where S is a singlet Higgs field. Derive the extended (F, D, and SSB) Higgs potential (for the doublets and singlet) in this case, and its minimization conditions. Show that the potential exhibits a discrete Z_3 symmetry broken only spontaneously by the Higgs *vev's*, which is the (cosmological) downfall of the model due to the associated domain wall problems at a post inflationary epoch. Write down the CP-even Higgs and neutralino extended mass matrices in this model. Find the tree-level light Higgs mass in the limit $\tan\beta = 1$ and show that it does not vanish (and that indeed the potential does not have a flat direction in this case). This is because a new term $\propto \sin^2 2\beta$ appears alongside the $\cos^2 2\beta$ term [107].

Summary

Though no evidence for supersymmetry has been detected as of the writing of this manuscript, indications in support of the framework have been accumulating and include the unification of gauge couplings; the heavy t-quark; and electroweak data preference of (i) a light Higgs boson and (ii) no "new-physics" quantum modifications to various observables. While individually each argument carries little weight, the combination of all indications is intriguing. The first three topics were explored in this part of the notes. We did so in a way that provided the reader with a critical view, as well as with a road map to many of the issues not discussed here in detail. The current status of the different models and proposals was emphasized.

The first issue, unification, also served as a warm up exercise in renormalization. The second issue provided the foundation to understanding the weak scale in the context of supersymmetry: Electroweak symmetry breaking, and hence the induction the weak scale, occurs dynamically and can be argued to be unique and to be based on perturbativity and renormalization rather than on ultraviolet details. The third issue addressed the perhaps most important aspect of any electroweak-scale model, the Higgs boson and its characteristics. Electroweak quantum corrections were investigated in detail in Ref. [117] and will not be discussed here. Useful reviews not mentioned so far include Refs. [118, 119].

A different set of relations between the infrared and ultraviolet, namely, the (very) low-energy constraints on the sparticle spectrum and couplings, will be discussed next in Part IV. In particular, we will demonstrate how low-energy constraints translate to ultraviolet principles.

Part IV

An Organizing Principle

9. The Flavor Problem

The most general MSSM Lagrangian is still described by many parameters. In particular, an arbitrary choice of parameters can lead to unacceptably large flavor changing neutral currents ("The Flavor Problem"), e.g. meson mixing and rare flavor-changing decays. Also, if large phases are present, unacceptable (either flavor conserving or violating) contributions to CP-violating amplitudes could arise ("The CP Problem"), e.g. the flavor conserving electron and neutron dipole moments.

Here, we focus on the flavor problem. The flavor-conserving CP problem [120] is, in practice, less constraining: It may be resolved independently if no new large physical phases are present in the weak-scale Lagrangian, i.e., they are absent in the high-energy theory due to a symmetry or coincidence (see the next chapter), or alternatively large relative phases could be diminished by renormalization effects and could be "renormalized away" [121]. It may even be resolved in some (very) special cases by cancellations [122], though tuning cancellations is at best equivalent to assuming small phases to begin with. (Nevertheless, the physical implications may be different.)

The flavor problem stems from the very basic supersymmetrization of the SM. Recall that even though each fermion flavor is coupled in the SM with an arbitrary 3×3 Yukawa (or equivalently, mass) matrix, the unitary diagonalization matrix guarantees that the fermion-fermion coupling to a neutral gauge boson is basis independent and flavor blind at tree-level. This is also true to a sfermion-sfermion coupling to a neutral gauge boson which is now multiplied by the sfermion unitary rotation matrices. However, by supersymmetry, there exists also a fermion-sfermion-gaugino vertex. If both fermion and sfermion mass matrices are diagonalized, then the gaugino vertex is rotated by two independent unitary matrices and, in general, their product $U_f U_{\tilde{f}}^{\dagger} \neq I$ is not trivial. The rotations maintain in this case some (non-trivial) flavor structure and the gaugino-fermion-sfermion vertex is not flavor diagonal. That is, generically the theory contains, for example, a $g_s \tilde{g} d \tilde{s}^*$ vertex! At the loop level, such new flavor changing couplings lead to new contributions to SM FCNC observables, just as flavor changing charged currents do within the SM.

The flavor problem, however, is a blessing in disguise: It provides one with an efficient organizing principle which eliminates some of the arbitrariness in the parameter space. We will apply this insight in the next chapter. Let us first proceed with some concrete examples which will reveal these principles.

9.1 Meson Mixing

In the SM, contributions to meson mixing from gauge-boson - quark couplings arise only at one loop. Even then, given the small mass differences among the light fermions and the smallness of light-heavy fermion mixing, such contributions, particularly to meson mixing, are highly suppressed (the GIM mechanism). However, given the flavor violating gaugino vertex in supersymmetry, it is straightforward to construct quantum corrections for example, to $K - \bar{K}$ mixing, which violate flavor by two units: See Fig. 9.1 for illustration. Such corrections could be *a priori* arbitrarily large. (Adding phases, new contributions to the SM parameter ϵ_K arise and provide a strong constraint.)

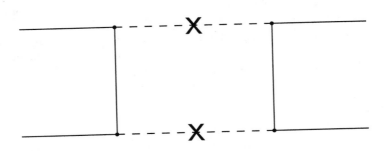

Fig. 9.1. A generic "box" diagram contribution to meson mixing with gauginos and squarks circulating in the loop. The Flavor changing is approximated here as flavor changing squark-mass insertions, which are denoted explicitly. In the case of kaon mixing, non-trivial phases would lead to a contribution to the parameter ϵ_K.

Such effects are more conveniently evaluated in the fermion mass basis in which flavor violation is encoded in the sfermion mass-squared matrix, which retains inter-generational off-diagonal entries,

$$
M_{\tilde{f}}^2 = \begin{pmatrix} m_{\tilde{f}_1}^2 & \Delta_{\tilde{f}_1 \tilde{f}_2} & \Delta_{\tilde{f}_1 \tilde{f}_3} \\ \Delta_{\tilde{f}_2 \tilde{f}_1} & m_{\tilde{f}_2}^2 & \Delta_{\tilde{f}_2 \tilde{f}_3} \\ \Delta_{\tilde{f}_3 \tilde{f}_1} & \Delta_{\tilde{f}_3 \tilde{f}_2} & m_{\tilde{f}_3}^2 \end{pmatrix}. \tag{9.1}
$$

In this basis the sfermions of a given sector $f = Q, U, D, L, E$ mix at tree level. The matrix (9.1) corresponds to the CKM rotated original sfermion

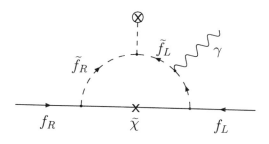

Fig. 9.2. One-loop radiative magnetic moment operator (with the chiral violation provided by the sfermion left-right mixing). $\widetilde{\chi}$ is a neutralino or chargino, and in the quark case there exists also a gluino loop.

mass-squared matrix ($\Delta_{\tilde{f}_i \tilde{f}_j} \propto m^2_{\tilde{f}_i} - m^2_{\tilde{f}_j}$). Additional flavor-violation arises, in principle, from the A-matrices. However, we ignore chirality labels, and left-right mixing (which would further complicate matters) is not included here.

When the above (insertion) approximation is applied to K meson mixing one has, for example, for the $K_L - K_S$ mass difference

$$\frac{\Delta m_K}{m_K f_K^2} \simeq \frac{2}{3} \frac{\alpha_3}{216} \frac{1}{\widetilde{m}^2} \left(\frac{\Delta_{\tilde{d}\tilde{s}}}{\widetilde{m}^2} \right)^2 , \tag{9.2}$$

where we used standard notation for the kaon mass m_K and for the relevant form factor f_K, and for simplicity dimensionless functions are omitted. This leads to the constraint on the ratio of a typical sqaurk mass $\widetilde{m}$ and the inter-generational mixing

$$\left(\frac{500 \text{ GeV}}{\widetilde{m}^2} \right) \left(\frac{\Delta_{\tilde{d}\tilde{s}}}{\widetilde{m}^2} \right) \lesssim 10^{-(2-3)}, \tag{9.3}$$

where the complete amplitude, including higher order corrections, was evaluated [123] in deriving (9.3). Weaker constraints are derived from B- and from D-meson mixing.

9.2 Magnetic and Electric Moments

Another set of observables, which is of particular importance, is given by magnetic moments with virtual sparticles circulating in the loop, shown in Fig. 9.2. SM contribution to any anomalous and transition magnetic moments is a one-loop effect. This is also the case for the supersymmetry contribution which is therefore of the same order of magnitude. Henceforth, magnetic moments could shed light on any new physics in general, and supersymmetry and its parameter space, in particular. Here, we review a few (experimentally) promising examples. Note that contributions to the various magnetic operators are strongly correlated within a given framework. (See the next chapter for frameworks.) Hence, a combined analysis provides a stronger probe than examining each observable individually. However, this is difficult to do in a model-independent fashion.

9.2.1 Flavor Conserving Moments

To begin with consider flavor conserving magnetic moments[1],

$$(eQ_f/2m_f)a_f \bar{f}\sigma_{\mu\nu}f F^{\mu\nu},$$

for example, the anomalous muon magnetic moment a_μ [124]. The flavor-conserving muon magnetic moment is currently probed far beyond current bounds at the Brookhaven E821 experiment, offering a rare opportunity for evidence of new physics [125]. Sparticle loops generically contribute

$$a_\mu^{\text{SUSY}} \sim \pm \frac{\alpha_2}{4\pi} \frac{m_\mu^2 M_Z \tan\beta}{\widetilde{m}^3}, \qquad (9.4)$$

where $\widetilde{m}$ is a typical superpartner (in this case, chargino) mass scale, and the sign is determined by the sign of the μ parameter. (In the special case of models in which the muon mass comes about radiatively rather from a tree-level Yukawa coupling one has [61] $a_\mu^{\text{SUSY}} \sim +m_\mu^2/3\widetilde{m}^2$, which is significantly enhanced.) This is a most interesting "flavor-conserving" $\tan\beta$-dependent constraint/prediciton in supersymmetry. In particular, as $\tan\beta$ increases, a measurement of $|a_\mu^{\text{SUSY}}| > 0$ can place a decreasing upper bound on the sparticle scale $\widetilde{m}$.

Similarly to the flavor conserving magnetic moment, the same loop diagram, but with an added phase[2], corresponds to a (flavor-conserving) dipole moment. For example, the electron dipole moment constraints slepton masses to be in the multi-TeV range if the corresponding "soft" phase ϕ (for example a gaugino or A-parameter phase, or their relative phase) is large,

$$\frac{\widetilde{m}}{\sqrt{(M_Z/\widetilde{m})\tan\beta\sin\phi}} \gtrsim 2 \text{ TeV}. \qquad (9.5)$$

---

[1] Note that the "pure fermionic" form of the operator, which dependes on the Pauli matrix combination $\sigma_{\mu\nu}$, suggests that it vanishes in the supersymmetric limit.

[2] If phases are present then the flavor violating parameter ϵ'_K/ϵ_K may also receive "soft" contributions. ϵ'_K/ϵ_K has been recently measured with some accuracy [7] and could actually accommodate new-physics contributions.

9.2.2 Transition Moments: $b \to s\gamma$

Returning to flavor violation, the same magnetic operator could connect two different fermion flavors leading to magnetic transition operators which violate flavor by one unit and where again the SM and supersymmetry contributions arise at the same order. The most publicized case is that of the $b \to s\gamma$ operator [126]; the first such operators whose amplitude was measured. The projected precision of the next generation of experiments is expected to allow one to disentangle SM from new-physics contributions. The generic supersymmetry contribution to the $b \to s\gamma$ branching ratio arises (just as in the SM and 2HDM) due to CKM rotations V. It reads

$$BR(b \to s\gamma) \simeq BR(b \to ce\bar{\nu}) \frac{|V_{ts}^* V_{tb}|^2}{|V_{cb}|^2} \frac{6\alpha}{\pi} \left[\mathrm{SM} + \mathrm{CH} \pm \frac{m_t^2 \tan\beta}{\tilde{m}^2} \right]^2 . (9.6)$$

The first contribution is the SM W-loop (written for reference). Currently, the SM term is consistent by itself with the experimental measurement, but the theoretical and experimental uncertainties accommodate possible new physics contributions. The second term is from a charged-Higgs loop. The third term is the one arises in supersymmetry from the SM CKM rotations which, by supersymmetry, apply also, e.g. to chargino vertices.

The charged Higgs contribution, also not written explicitly, is positive, and comparable to the SM contribution for a relatively light charged Higgs $m_{H^\pm} \lesssim$ 200-300 GeV (*i.e.*, below the higgs decoupling limit of the previous chapter). Therefore, in the heavy sparticle limit, in which sparticle (in particular, squark) contributions decouple, one can constrain (already using current data [7]) the charged Higgs mass from below. This is the case, for example, in the gauge mediation framework which is discussed in the next section.

The third term is present in any supersymmetric theory. Its sign is given by the relative sign of A_t and μ, and $\tilde{m}$ is of the order of the chargino and stop masses. There could also be a contribution in supersymmetry which is intrinsically flavor violating (see Ref. [126] and Gabbiani *et al.* in Ref. [127]),

$$BR(b \to s\gamma) \simeq \frac{2\alpha_3^2 \alpha}{81\pi^2} \mathcal{F} \frac{m_b^3 \tau_B}{\tilde{m}^2} \left(\frac{\Delta_{\tilde{s}\tilde{b}}}{\tilde{m}^2} \right)^2 \qquad (9.7)$$

with $\mathcal{F} \simeq \mathcal{O}(1 - 10)$, $\tau_B \sim 10^{-12}$ sec the B-meson mean life time, and $\tilde{m}$ of the order of the gluino and sbottom masses. We observe two quite different contributions: Eq. (9.6) with flavor violation induced by SM CKM (fermion) rotations of an otherwise flavor conserving chargino-quark-squark vertex, for example; and eq. (9.7) with explicit flavor violation (described by the sfermion mass-squared matrix (9.1)). (It is possible that the two contributions cancel out to some degree.)

While $b \to s\gamma$ has been observed and the branching ratio measured, it does not yet provide a signal of new physics, nor it provides strong constraints, in particular, if both contributions (9.6) (including the charged-Higgs term) and (9.7) are considered. Nevertheless, some constraints on relevant parameters can be obtained (as a function of other parameters), which is useful when analyzing specific models. This situation may improve in the near future.

9.2.3 Transition Moments: $\mu \to e\gamma$

Moderate to strong constraints arise in the most interesting case of individual-lepton number (lepton-flavor) violating (but total lepton-number conserving) magnetic transition operators such as $\mu \to e\gamma$. These operators vanish in the SM in which lepton number is conserved. Thus, an observation of such a process will provide a clear indication for new physics!

In supersymmetry slepton mass-squared matrices and trilinear couplings may not respect these accidental flavor symmetries of the SM, allowing for such processes. Lepton-flavor violations could further be related to specific models of total lepton number violation and neutrino masses and/or grand-unified models (see Chap. 11) where quarks and leptons are predicted to have similar flavor structure and rotations at high energies.

LFV experiments are therefore of extreme importance, in particular, given that lepton-flavor violation in (atmospheric) neutrino oscillations was established experimentally [4]. However, the experiments are also extremely difficult given the small amplitudes expected for such processes (which are not enhanced by large QCD or Yukawa couplings). In the case of $\mu \to e\gamma$, one currently has

$$\widetilde{m} \gtrsim 600 \text{ GeV} \sqrt{\left(\frac{M_Z}{\widetilde{m}}\right)\left(\frac{\Delta_{\tilde{e}\tilde{\mu}}}{\widetilde{m}^2}\right)} \tan\beta, \qquad (9.8)$$

where $\widetilde{m}$ here is of the order of the slepton mass.

9.3 From Constraints to Principles

Flavor violations (and magnetic moments) in supersymmetry and the corresponding signals and constraints have been studied by many authors, and a sample of papers is listed in Ref. [127]. Supersymmetry is not the only extension of the SM which is constrained by flavor (and CP) conservation. In fact, such constraints are generic to any extension since by its definition new-physics extends the SM unique structure that eliminates tree-level FCNC's – a property which is difficult to maintain. Nevertheless, unlike strongly-coupled theories or theories with extra dimensions (whether supersymmetric or not), in the case of perturbative supersymmetry the theory is well-defined and calculable, and hence, the problem is well-defined and calculable (as illustrated above), leaving no place to hide.

The satisfaction of the above and similar constraints derived from low-energy observables leads to the consideration of a small number of families of special models, and hence, provides an organizing principle as well as restores predictive power. Clearly, realistic models require nearly flavor-conserving gaugino couplings. Models with the right balance of the universality limit $\Delta_{\tilde{f}_i\tilde{f}_j} \to 0$ (in the physical fermion mass basis) and the decoupling limit $\widetilde{m} \gg M_Z$ (in which the sfermions and their couplings can be integrated out) can achieve that. The former limit is difficult to justify without a governing principle (though such a principle may be a derivative of the ultraviolet theory and only seem as arbitrary from the low-energy point of view); the latter limit may seem naively as contradictory to the notion of low-energy supersymmetry and its solution to the hierarchy problem. Nevertheless, these are the keys for the resolution of the flavor problem.

We continue in the next chapter with a brief review of the realization of these limits in various frameworks for the origins of the SSB parameters.

Exercises

9.1 Find $\Delta_{\tilde{f}_i\tilde{f}_j}$ (eq. (9.1)) by rotating the diagonalized sfermion mass-squared matrix to the fermion mass basis. Generalize eq. (9.1) to include left-right mixing.

9.2 Write tree-level meson mixing diagrams in theories with R-parity violation. Find a simple condition on the R-parity violating Yukawa couplings under which such tree-level contributions vanish.

9.3 The magnetic moment loop is nearly identical to the flavor conserving one-loop correction to a fermion mass (with sparticles circulating in the loops). Write and estimate these loops. Show that the leading QCD correction to the b-mass $\sim y_b(2\alpha_3/3\pi)\langle H_2^0\rangle \sim m_b \tan\beta(2\alpha_3/3\pi)$ if all SSB (and μ) parameters are equal. (At what scale are the different parameters evaluated?) In realistic models the correction is $\sim \pm 2\% m_b \tan\beta$, as alluded to in Sect. 6.2.1. What is the correction to the top mass in this limit? Why are the corrections to lepton masses highly suppressed?

9.4 Calculate the numerical coefficients of the CKM and intrinsic supersymmetry contributions to $BR(b \to s\gamma)$. (Use Ref. [7] for data inputs.) Add appropriately in quadrature and (assuming a heavy charged Higgs boson) constrain $\Delta_{\tilde{s}\tilde{b}}/\widetilde{m}^2$ as a function of $\tan\beta/\widetilde{m}^2$, and vice versa. (Assume similar values for the different $\widetilde{m}^2$'s.)

9.5 Assume that at the GUT scale leptons and quarks are rotated by the same CKM rotations, and that the only other source of LFV is $m_{\tilde{E}_3}^2 < m_{\tilde{E}_1}^2 = m_{\tilde{E}_2}^2$. Estimate the induced contributions to the $\mu \to e\gamma$ branching ratio. What parameters does it depend on?

10. Models

We are now in position to use the flavor problem to organize the SSB parameter space. First, however, the parameter space itself needs to be efficiently parameterized. We begin by postulating that supersymmetry is broken spontaneously in some "hidden" sector of the theory which contains the Goldstino and which interacts with the SM "observable" sector only via a specific agent. The agent is then the messenger of supersymmetry breaking. These messenger interactions cannot be renormalizable tree-level interactions or otherwise the SM could couple directly to the Goldstino multiplet and one would encounter many of the problems that plagued our attempt in Sect. 5.4 to break supersymmetry spontaneously in the SM sector.

The messengers and their interactions then decouple at a scale $\Lambda_{\text{mediation}}$ which effectively serves the scale of the mediation of the SSB parameters in the low-energy theory, i.e., this is the scale below which these parameters can be considered as "hard" and treated with the renormalization-group formalism (as we did in Chap. 7 in our discussion of radiative symmetry breaking). This general "hidden-messenger-observable" phenomenological framework describes most of the models and will suffice for our purposes. It is the details of the messenger interactions and the mediation scale that impact the spectrum parameters and the flavor problem and its solution. (It is important to recall that gauge, Yukawa, and quartic interactions are dictated by supersymmetry at all energies above the actual sparticle mass scale, and that only the dimensionful parameters interest us in this section.)

10.1 Operator Expansion

In order to review the different classes of models, let us re-write the soft parameters (and for completeness, also the μ-parameter) in a generalized operator form, using the tools of (global) supersymmetry described in Chap. 4. We can express the most general set of operators as an expansion in powers of $\sqrt{F_{X,Z}}$, where X and Z are singlet and non-singlet superfields which spontaneously break (or parameterize such breaking) supersymmetry. They contribute (if not provide) to the Goldstino component of the physical gravitino; their F-components are order parameters of supersymmetry breaking as seen in the SM sector. The operators are suppressed by powers of

$M \simeq \mathcal{O}(\Lambda_{\text{mediation}})$. The fields X and Z which couple to the SM sector are the messengers of supersymmetry breaking, and the interaction which couple them to the SM sector is its agent.

The gravitino mass can be shown (by the local-supersymmetry condition of a vanishing cosmological constant) to be $m_{3/2} = \sum_i F_i/\sqrt{3}M_P$ where M_P is as before the local-supersymmetry expansion parameter: The reduced Planck mass $M_P = M_{\text{Planck}}/\sqrt{8\pi} \simeq 2.4 \times 10^{18}$ GeV. Otherwise no direct application of local supersymmetry is needed.

The leading terms in this expansion (see Chap. 13 for a generalization) which generate the μ parameter and soft terms have the following form:

$$\text{Scalar masses}: \quad a \int d^2\theta d^2\bar{\theta}\, \Phi_i^\dagger \Phi_i \left[\frac{X^\dagger X}{M^2} + \frac{Z^\dagger Z}{M^2} + \cdots\right] \quad (10.1)$$

$$\mu \text{ parameter}: \quad b \int d^2\theta d^2\bar{\theta}\, H_1 H_2 \left[\frac{X^\dagger}{M} + h.c. + \cdots\right]$$

$$\text{and} \quad c \int d^2\theta\, H_1 H_2 \left[\frac{X^n}{M^{n-1}}\right.$$

$$\left. +c'\frac{\hat{W}^\alpha \hat{W}_\alpha}{M^2} + h.c. + \cdots\right] \quad (10.2)$$

$$\text{Higgs mixing } (m_3^2): \quad d \int d^2\theta d^2\bar{\theta}\, H_1 H_2 \left[\frac{X^\dagger X}{M^2} + \frac{Z^\dagger Z}{M^2} + \cdots\right] \quad (10.3)$$

$$\text{Gaugino masses}: \quad e \int d^2\theta\, W^\alpha W_\alpha \left[\frac{X}{M} + h.c. + \cdots\right] \quad (10.4)$$

$$A-\text{terms}: \quad f \int d^2\theta\, \Phi_i \Phi_j \Phi_k \left[\frac{X}{M} + h.c. + \cdots\right]$$

$$\text{and} \quad g \int d^2\theta d^2\bar{\theta}\, \Phi_i^\dagger \Phi_i \left[\frac{X}{M} + h.c. + \cdots\right], \quad (10.5)$$

where $a-g$ are dimensionless coefficients, the W_α ($\hat{W}_\alpha$) are spinorial gauge supermultiplets containing the standard model gauginos (a hidden-sector gauge field which condenses), the Φ_i are SM (observable) chiral superfields, and X and Z represent supersymmetry-breaking gauge singlet and non-singlet superfields, respectively. These terms give the SSB parameters and the μ parameter when the X and Z fields acquire F-term vev's: $X \to \theta^2 F_X$, $Z \to \theta^2 F_Z$ (and, in the second source for A-terms, $\Phi_i^\dagger \to \bar{\theta}^2 F_{\Phi_i}^* \sim \bar{\theta}^2 \Phi_j \Phi_k$).

The μ-parameters could also arise from a supersymmetry conserving vev of x, $X \to x + \theta^2 F_X$ The case $n = 1$ in eq. (10.2) corresponds to the NMSSM and one has to arrange $\langle x \rangle \simeq M_{\text{Weak}} \simeq m_{\text{SSB}}$. In this case $m_3^2 \sim A\langle x \rangle \sim A\mu$. The case $n = 2$ corresponds to $\langle x \rangle \simeq \sqrt{M M_{\text{Weak}}}$, which coincides with the spontaneous supersymmetry breaking scale in supergravity (see the next section). The case $n = 3$ has $X^3 \equiv \langle W_{\text{hidden}} \rangle$, where $\langle W_{\text{hidden}} \rangle$ is a hidden superpotential vev which (in local supersymmetry) also breaks supersymmetry. It can be related to first operator in (10.2) via redefinitions (a Kähler transformation in local supersymmetry).

Observe that if only one singlet field participates in supersymmetry breaking then the SSB parameters have only one common phase, and at most one additional phase arises in the μ parameter. Since it can be shown that two phases can always be rotated away and are not physical, there is no "CP problem" in this case.

10.2 Supergravity

It is widely thought that if supersymmetry is realized in nature then it is a local (super)symmetry, *i.e.*, a supergravity (SUGRA) theory. (In particular, in this case the cosmological constant is not an order parameter and could be fine-tuned to zero.) Supergravity interactions provide in this case the theory of (quantum) gravity at some ultraviolet scale (though it is a non-renoemalizable theory and hence is probably not the ultimate theory of gravity). As a theory of gravity, supergravity interacts with all sectors of the theory and hence provides an ideal agent of supersymmetry breaking. In fact, even if other agents/messengers exist, supergravity mediation is always operative once supersymmetry is gauged. Its mediation scale is, however, always the reduced Planck mass. From the operator equations (10.1) – (10.5) and from $m_{3/2} \sim F/M_P$ one observes that SUGRA contribution to the SSB parameters is always $\sim m_{3/2}$, so it could be the leading contribution for a sufficiently heavy gravitino $m_{3/2} \gtrsim M_Z$, but otherwise its contributions are suppressed. It may be argued that the hierarchy problem is then the problem of fixing the gravitino mass (which may be solved by dynamically breaking supersymmetry in the hidden sector).

Let us then consider the case of a "heavy" gravitino, $m_{3/2} \simeq \mathcal{O}(100-1000)$ GeV (and $\mathcal{O}(1)$ coefficients). Supersymmetry is spontaneously broken in this case at a scale $\sqrt{F} \sim \sqrt{m_{3/2}M_P} \sim 10^{10-11}$ GeV. The scale of the mediation is always fixed in supergravity to M_P. The hidden sector is truly hidden in this case in the sense that it interacts only gravitationally with the observable sector (though supergravity interactions could take various forms). Supergravity mediation leads to some obvious observations. First, if $\mu = 0$ in the ultraviolet theory, it is induced (the first term in eq. (10.2)) once supersymmetry is broken and is of the order of the gravitino mass, resolving the general μ-problem [101, 102, 103]. (More solutions arise if the model is such that $\langle x \rangle \simeq \sqrt{F}$ or supersymmetry is broken by gaugino condensation in the hidden sector [54, 128] $\langle \hat{W}^\alpha \hat{W}_\alpha \rangle \simeq \langle F \rangle^{3/2}$.) Secondly, since the gravitino mass determines the weak scale $m_{\tilde{f}} \simeq m_{3/2} \simeq M_Z$, the decoupling limit cannot be realized. Hence, in order for the SSB parameters to conserve flavor, supergravity has to conserve flavor so that the universality limit can be realized.

Naively, gravity is flavor blind. In general, however, supergravity (and string theory) is not! Re-writing, for example, the quadratic operators with flavor indices one has $a_{\alpha\beta ab}Z_\alpha Z_\beta^\dagger \Phi_a \Phi_b^\dagger$. The universality limit corresponds to the special case of $a_{\alpha\beta ab} = \hat{a}_{\alpha\beta}\delta_{ab}$, where at least each flavor sector is described by a unique $\hat{a}_{\alpha\beta}\delta_{ab}$ coefficient. (Different flavor sectors could be distinguished by an overall factor.) In order to maintain universality one has to forbid any other hidden-observable mixing such as $(ZZ^\dagger)(\Phi\Phi^\dagger)^2/M^4$ which lead to quadratically divergent corrections (proportional to the gravitino mass). The latter, in turn, spoil the universality at the percentile level [129, 130, 128] (which is the experimental sensitivity level), $a_{\alpha\beta ab} = \hat{a}_{\alpha\beta}(\delta_{ab} + (N_{ab}/16\pi^2))$ for some flavor-dependent counting factor N_{ab}. It was argued that the universality may be a result of a flavor (contentious or discrete) symmetry which is respected by the hidden sector and supergravity or a result of string theory. If the symmetry is exact (which most likely requires it to be gauged) then universality can hold to all orders. If universality is a result of a "string miracle" (such as supersymmetry breaking by a stabilized dilaton [131]) then it is not expected to be exact beyond the leading order.

A minimal assumption that is often made as a "best first guess" is that of total universality (of scalar masses squared), A-term proportionality (to Yukawa couplings), and gaugino mass unification, which lead to only four ultraviolet parameters (in addition to possible phases and the sign of the μ-parameter): $m_\phi^2(\Lambda_{\rm UV}) = m_0^2$ (universality), $A_{ijk}(\Lambda_{\rm UV}) = A_0 y_{ijk}$ (universality and proportionality), $M_i(\Lambda_{\rm UV}) = M_{1/2}$ (gaugino unification), $m_3^2(\Lambda_{\rm UV}) = B_0\mu(\Lambda_{\rm UV})$, which are all of the order of the gravitino mass [132]. (μ is not a free parameter in this case but is fixed by the M_Z constraint, eq. (8.3), only its sign is a free parameter.) This framework is sometimes called minimal supergravity. It may be that only the operator (10.4) is present at the mediation scale ($m_0 = A_0 = 0$) and that all other SSB parameters are induced radiatively by gaugino loops. (For example, see Kelley et al. in Ref. [100].) In this case, the gauge interactions guarantee universality in 3×3 subspaces: GIM universality. Such models are highly predictive. Universality (whether in each 3×3 subspace or for all fields) is the mystery of supergravity models, but since the mediation itself and the μ-parameter are trivially given in this framework, it cannot be discounted and the price may be worth paying.

It was also proposed that supergravity mediation is carried out only at the quantum level [133] (anomaly mediation) via supergravity quantum corrections which generically appear in the theory. The relevant coefficients are then given by loop factors with specific pre-factors which are determined by the low-energy theory, e.g. $e \sim b_i g_i^2/16\pi^2$ is given by the low-energy (i.e., weak-scale) one-loop β-function (and there is no gaugino mass unifcation). These proposals, though economic and elegant and with unique signatures (the winos are lighter than the bino, for example) face difficulties in deriving a consistent scalar spectrum and the μ-parameter. In particular, the coefficients for the sfermions squared masses are given by the respective two-loop anomalous dimensions, which though universal (up to Yukawa-coupling corrections) are negative in the case of the sleptons (aside from the $\tilde{\tau}$ if y_τ is sufficiently large). This scenario therefore predicts negative squared masses for sleptons. Known cures are contrived and/or reintroduce ultraviolet dependencies. A lesson from these proposals, however, is that gaugino unification cannot be exact. (For a general study of corrections to gaugino unification, see Ref. [134].) More generally, quantum mediation could extend beyond anomaly mediation [135, 136] and could lead to viable models.

10.3 Gauge Mediation

It may be that supergravity mediation is sufficiently suppressed by a light gravitino mass (and hence no assumption on its flavor structure is needed). Then, a new mediation mechanism and messenger sector are required. An attractive option is that the messenger interactions are gauge interactions so that universality is an automatic consequence (e.g. gaugino mediation and anomaly mediation in SUGRA). This is the gauge-mediation (GM) framework. The hidden sector (which is not truly hidden now) communicates via, e.g. new (messenger) gauge interactions with a messenger sector, which, in turn, communicates via the ordinary gauge interactions with the observable sector.

It is sufficient to postulate that the new messenger gauge and Yukawa interactions mediate the supersymmety breaking to a (SM) singlet messenger $X = x + \theta^2 F_X$, which parameterizes the supersymmetry breaking in the messenger sector. (Some other hidden fields, however, with $F > F_X$ may dominate the massive gravitino.) The singlet X interacts also with SM non-singlet messenger fields V and $\bar{V}$. The Yukawa interaction $yXV\bar{V}$ in turn communicates the supersymmetry breaking to the messengers V and $\bar{V}$ via the mass matrix,

$$M_{v\bar{v}}^2 \sim \begin{pmatrix} y^2 x^2 & yF_X \\ yF_X^* & y^2 x^2 \end{pmatrix}. \tag{10.6}$$

In turn, the vector-like pair V and $\bar{V}$, which transforms under the SM gauge group (for example, they transform as 5 and $\bar{5}$ of $SU(5)$, $i.e.$, as down singlets and lepton doublets and their complex conjugates), communicates the supersymmetry breaking to the ordinary MSSM fields via gauge loops. The gauge loops commute with flavor and, thus, the spectrum is charge dependent but flavor diagonal, if one ensures that all other possible contributions to the soft spectrum are absent or are strongly suppressed. This leads to (GIM-)universality. Such models are often referred to as "gauge mediation of supersymmetry breaking" or messenger models [137].

This scenario is conveniently described by the above operator equations with $M \simeq \langle x \rangle$, $e \sim \alpha_i/4\pi$ a generic one-loop factor and $a \sim (\alpha_i/4\pi)^2$ a two-loop factor. (If the messenger dynamics itself is non-perturbative, by dimensionless analysis, $e \sim \alpha_i$ and $a \sim (\alpha_i)^2$ [138].) All other coefficients generically equal zero at the messenger scale. The sparticle spectrum, and hence, the weak scale, are given in this framework by $M_Z \sim m_{\mathrm{SSB}} \sim (\alpha_i/4\pi)(F_X/x)$, where α_i is the relevant SM gauge coupling at the scale $\Lambda_{\mathrm{mediation}} \sim x$. One then has $F_x/x \sim (4\pi/\alpha_i)M_Z \sim 10^5$ GeV. In the minimal version one assumes $\Lambda_{\mathrm{mediation}} \sim F_x/x \sim 10^5$ GeV; a similar scale for the spontaneous supersymmetry breaking in the hidden sector; as well as $F_X \sim x^2$; resulting in an one-scale model. The latter assumption could be relaxed as long as the phenomenologically determined ratio $F_x/x \sim (4\pi/\alpha)M_Z \sim 10^5$ GeV remains fixed. In general, $\Lambda_{\mathrm{mediation}} \sim x$ could be at a much higher scale [139]. Also, the scale of spontaneous supersymmetry breaking in the hidden sector could be one or two orders of magnitude higher than $\sqrt{F_X}$, for example, if F_X is induced radiatively by a much larger F-term of a hidden field (in which case X contributes negligibly to the massive gravitino).

The messengers induce gaugino masses at one-loop ($e \sim \alpha_i/4\pi$) and scalar masses at two-loops ($a \sim (\alpha_i/4\pi)^2$). (The messengers obviously cannot couple via gauge interactions to the other chiral fields at one-loop.) This leads to the desired relation $m_{\tilde{f}}^2 \sim M_\lambda^2$ between scalar and gaugino masses. In addition, there is a mass hierarchy $\sim \alpha_3/\alpha_2/\alpha_1$ between the heavier strongly interacting sparticles and the only weakly interacting sparticles which are lighter. (In detail, it depends on the charges of the messengers). In particular, gaugino mass relations reproduce those of gaugino mass unification. The A-parameters arise only via mixed Yukawa-gauge renormalization and are of no concern (since renormalization-induced A-parameters are proportional to Yukawa coupling). However, the Higgs mixing parameters also do not arise from gauge interactions, the Achilles heal of the framework.

One may introduce new hidden-observable Yukawa interactions for the purpose of generating μ and m_3^2, but then generically $b \simeq d \simeq (y^2/16\pi^2)$ for some generic Yukawa coupling y, leading to a hierarchy problem $|m_3^2| \sim |\mu|\Lambda_{\mathrm{mediation}}$. This overshadows the otherwise success of gauge mediation, and the possible resolutions [140] are somewhat technically involved and will not be presented here. It was also shown that the resulting uncertainty in the nature of the Higgs-messenger interactions introduces important corrections to the Higgs potential and mass [78].

Note that this framework has a small number of "ultraviolet" parameters and it is highly predictive (less so in extended versions). The gravitino is very light and is the LSP (with signatures such as neutralino decays to an energetic photon and Goldstino missing energy or a charged slepton escaping the detector, decaying only outside the detector to a lepton and a Goldstino). Also, the SSB parameters are not "hard" at higher scales such as the unification scale. Radiative symmetry breaking relies in this case on mass hierarchy mentioned above, $m_{\tilde{q}}^2(\Lambda_{\mathrm{mediation}}) \sim (\alpha_3/4\pi)^2 \Lambda_{\mathrm{mediation}}^2 \gg m_{H_i}^2(\Lambda_{\mathrm{mediation}}) \sim$

$(\alpha_2/4\pi)^2 \Lambda^2_{\text{mediation}}$, rather than on a large logarithm, and its solutions take a slightly different form than eqs. (7.6).

10.4 Hybrid Models I

Can the gauge-mediation order parameter F_X/x be induced using only supergravity interactions? (Supergravity serves in such a case only as a trigger for the generation of the SSB parameters.) The answer is positive.

An operator of the form

$$\int d^2\theta d^2\bar{\theta}\, X \left[\frac{Z^\dagger Z \Phi^\dagger \Phi}{M_P^3} + h.c. + \cdots\right] \tag{10.7}$$

leads to a quadratically divergent tadpole (see Fig. 3.1) loop (with Φ circulating in the loop) which is cut-off at M_P, and hence to a scalar potential of the form $V(x) = (|F_Z|^2/M_P)x + |\partial W/\partial X|^2 \sim m_{3/2}^2 M_P x + x^4$ (where we omit loop and counting factors and in the last step we assumed $W(X) = X^3/3$).

This is only a sketch of this hybrid frameork [129] in which a supergravity linear term in a singlet X can trigger a gauge mediation framework for $\sqrt{F_Z} \sim 10^{8\pm1}$ GeV (leading to $x \sim 10^5$ GeV and $F_X \sim x^2$, as in minimal gauge mediation). The clear benefit is that the hidden sector (Z in this case) remains hidden while X, which couples as usual $XV\bar{V}$, is a true observable sector singlet field. This leads to a simpler radiative structure and to a more stable model. The triggering gravity mediation (leading to the linear term) is carried out, as in anomaly mediation, only at the quantum level (but with a very different source than in anomaly mediation).

10.5 Hybrid Models II

A different hybrid approach is that of superheavy supersymmetry (or the $2-1$) framework, which as implied by its name, relies on decoupling in order to weaken the universality constraint. The conflict between naturalness and experimental constraints is resolved in this case by observing that, roughly speaking, naturalness restricts the masses of scalars with large Yukawa couplings, while experiment constrains the masses of scalars with small Yukawa couplings [141]. Naturalness affects particles which are strongly coupled to the Higgs sector, while experimental constraints are strongest in sectors with light fermions which are produced in abundance. This suggests that naturalness and experimental constraints may be simultaneously satisfied by an "inverted hierarchy" approach, in which light fermions have heavy superpartners, and heavy (third family) fermions have "light" $\mathcal{O}(M_Z)$ superpartners (hence, $2-1$ framework). Therefore, the third generation scalars (and Higgs) with masses $m_{\text{light}} \lesssim 1$ TeV satisfy naturalness constraints, while first and second generation scalars at some much higher scale m_heavy avoid many experimental difficulties.

A number of possibilities have been proposed to dynamically generate scalar masses at two hierarchically separated scales. Usually one assumes that $m_\text{light} \simeq m_{3/2}$ is generated by the usual supergravity mediation (and μ generation may follow SUGRA as well), while m_heavy is generated by a different "more important" (in terms of the relative contribution) mechanism which, however, discriminates among the generations. (Naturalness allows the stau, in some cases, to be heavy.)

Such a "more important" mechanism may arise from the D-terms of a ultraviolet anomalous flavor $U(1)$ (we did not discuss in these notes the case of an anomalous $U(1)$ [142]) with a non-vanishing D-term, $F_Z F_Z^\dagger \rightarrow \langle D \rangle^2$ in eq. (10.1), and $\langle D \rangle / M \gg m_{3/2}$ has to be arranged [143]. Alternatively, there could be a flavor (or horizontal) messenger mechanism at some intermediate energies [129, 144] (and in this case the gauged flavor symmetry is anomaly free and it is broken only at intermediate energies). The messenger model in this case follows our discussion above only that the messengers are SM singlets charged only under the flavor (horizontal) symmetry. In both of these examples a gauge horizontal (flavor) symmetry discriminates among the generations and does not affect the MSSM gaugino masses. For example, the horizontally neutral third generation sfermions and Higgs fields do not couple to the horizontal D-term (the first case) or to the horizontal messengers (the second case). The coefficients a in eq. (10.1) are in these cases flavor dependent but do not mix the light and heavy sfermions, a mixing which is protected by the horizontal gauge symmetry.

A fundamentally different approach [97] is that $F_Z \gg F_X$ (a limit which realizes an effective $U(1)_R$ symmetry in the low-energy theory) and all the boundary conditions for all the scalars are in the multi-TeV range (while gauginos are much lighter). The light stop squarks and Higgs fields, for example, are then driven radiatively and asymptotically to m_light. Indeed, and not surprisingly, one find that in the presence of large Yukawa couplings there is such a zero fixed point which requires, however, that the respective sfermion and Higgs boundary conditions have specific ratios (i.e., it is realized along a specific direction in field space as pointed out in Chap. 7). Here, the hierarchy is indeed inverted as the light scale is reached via renormalization by large Yukawa couplings.

In practice, all of these solutions are constrained by higher-order terms which couple the light and heavy fields and which are proportional to small Yukawa couplings or which arise only at two-loops. Such terms are generically suppressed, but are now enhanced by the heavy sfermions. The importance of such effects is highly model dependent and they constrain each of the possible realization in a different fashion, leaving more than sufficient room for model building.

Even though such a realization of supersymmetry would leave some sfermions beyond the kinematic reach of the next generation of collider experiments, some sfermion and gauginos should be discovered. It was noted that by measuring the ratio of the gaugino-fermion-(light)sfermion coupling and the gauge coupling, $g_{\tilde\chi f \tilde f_*}/g$ (the superoblique parameters [145]), one can confirm in many cases the presence of the heavy states by measuring logarithmically-divergent quantum corrections to these ratios $\sim \ln m_{\text{heavy}}$, providing an handle on these and other models with multi-TeV fields (e.g., the squarks in GM). The phenomenology of the $2-1$ approach was examined recently in some detail in Ref. [146].

10.6 Alignment

Finally, it was proposed that the Yukawa and sfermion squared mass (and trilinear A) matrices are aligned in field space, and hence, are diagonalized simultaneously [147]. Thus, $\Delta_{ij} = 0$ without universality. Such an alignment may arise dynamically once the (Coleman-Weinberg) effective potential is minimized with respect to some low-energy moduli X and Z. However, such a mechanism tends to be unstable with regard to higher-order corrections. Alternatively, it could be that such alignment is a result of some high-energy symmetry principle. Realization of the latter idea tend be cumbersome, in contrast to the simplicity of the assumption, and it is difficult to envision conclusive tests of specific symmetries (though the alignment idea itself may be ruled out, in principle, in sfermion oscillations [148] which cannot arise for $\Delta_{ij} \equiv 0$.) Nevertheless, this is another possibility in which case F_X/M and its square are aligned with the Yukawa matrices and are diagonal (only) in the physical mass basis of fermions.

Exercises

10.1 Integrate the operator equations (10.1)–(10.5) to find the SSB and μ-parameters.

10.2 Solve for the sfermion and Higgs mass-squared parameters in the case that supergravity mediation generates only the gaugino mass. (See eq. (7.6).) What is the LSP (assume $|\mu| \simeq M_3$), and is it charged or neutral? Assume instead complete universality at the supergravity scale $m_{\tilde f}^2 = m_H^2 = m_0^2$ for all sfermion and Higgs fields. In what fashion gaugino masses and Yukawa couplings break universality? Why is the breakdown of univesality by corrections proportional to Yukawa couplings "relatively safe" (as far as FCNC are concerned)?

10.3 In the spirit of the previous exercise, show that squark mass squared matrices are never truly universal. What are the implications, for example, for gluino couplings? Flavor-changing gluino couplings are important for meson mixing, $b \to s\gamma$, and proton decay amplitudes.

10.4 Supergravity mediation in the presence of a grand unified theory induces the SSB parameters for that theory, not for the MSSM. How many independent soft parameters are in this case at the unification scale (with and without universality) for minimal $SU(5)$; $SO(10)$? Renormalization within a grand-unified epoch introduces radiative corrections to the stau squared mass ($m^2_{\tau_L}$ or $m^2_{\tau_R}$?) which are proportional to the large top-Yukawa coupling. The corrections break universality in the slepton sector. The breakdown of slepton universality induces, in turn, contributions to low-energy lepton flavor violation such as the $\mu \to e\gamma$ process (Ex. 9.5). For sufficiently light sleptons ($m_{\tilde{l}} \lesssim 300\text{-}400$ GeV) it may be observable in the next generation of the relevant experiments, if built.

10.5 Show that the winos are heavier than the bino in anomaly mediation (as defined in the text). Consider the chargino-LSP mass difference, assuming both are gaugino-like.

10.6 Derive the messenger mass matrix (10.6).

10.7 Show that in the minimal gauge mediation model described above $e = (\alpha_i/4\pi)T_i$ for gaugino i, and normalizing Dynkin index T_i to unity (e.g. for **5** and **5̄** of $SU(5)$ messengers) that for each sfermion (and Higgs) $a = 2\sum_i(\alpha_i/4\pi)^2 C_i$ where $C_1 = (3/5)Y^2$, $C_2 = 3/4$ for an $SU(2)$ doublet and $C_3 = 4/3$ for an $SU(3)$ triplet are the Casimir operators.

10.8 Derive the relation $|m_3^2| \sim |\mu|\Lambda_{\text{mediation}}$ in gauge mediation.

10.9 Calculate F_X in the hybrid supergravity - gauge-mediation model. Calculate the gravitino mass in this and in the "traditional" gauge-mediation models and show that supergravity effects $\sim m_{3/2}^2$ are indeed negligible in both cases.

10.10 Compare typical sparticle mass patterns in minimal SUGRA and in minimal GM.

Summary

The electroweak/Higgs scale was already (technically) explained, and in a fairly model-independent way, by the discussion in Part III. Nevertheless, its magnitude depends on at least two free parameters; the scale of spontaneous supersymmetry breaking $\sqrt{F}$ and the scale of its mediation to the SM fields M. The two scales combine to give the SSB scale $\sim F/M$. Henceforth, their ratio is fixed by the electroweak scale constraint. On the other hand, the scales of supersymmetry breaking and of its mediation relate to the ultraviolet origins of the soft terms in the fundamental theory, and are fixed by the dynamics of a given framework. This provides a useful linkage between the infrared and the ultraviolet.

In this part of the notes we parameterized and expanded the ultraviolet theory in terms of these scales and discussed different realizations, including various variants of supergravity and gauge-mediation models. In doing so we related some of the finer details of the infrared and ultraviolet physics. In order to perform this exercise, however, we first had to search for an organizing principle applicable to the vast SSB "model space".

Indeed, the most general MSSM Lagrangian is described by many arbitrary parameters, in particular, once flavor and CP conservation are not imposed on the soft parameters. Aside from the obvious loss of predictive power, many generic models are actually in an apparent conflict with the low-energy data: An arbitrary choice of parameters can result in unacceptably large FCNC ("The Flavor Problem") and, if large phases are present, also large (either flavor conserving or violating) contributions to CP-violating amplitudes ("The CP Problem"). Clearly, one has to identify those special cases which evade these constraints, and by doing so the whole framework regains predictive power. We therefore turned in our search of an organizing principle to the question of lack of low-energy evidence for supersymmetry.

The absence of observable contributions to FCNC was shown to indeed provide the desired organizing principle for high-energy frameworks, which themselves attempt to organize the model many parameters. Understanding the solution of the flavor problem (which is straightforward in the SM, given is special structure, but not in any of its extensions) goes to the heart of the question of origins of the theory. Though the number of options was narrowed down, the origin of the SSB parameters remains elusive. Discovery of supersymmetry will open the door to probing the mediation scale and mechanism, and hence, to a whole new (ultraviolet) arena.

In the remaining part of these notes we will return to and expand on a small number of topics. (The selected topics include a generalization of the operator classification given in this part of the notes.)

Part V

Selected Topics

11. Neutrinos

The confirmation of neutrino oscillations [4] provides a concrete and first proof of physics beyond the SM. If nature realizes supersymmetry, an option which we explore in this manuscript, then neutrino mass and mixing (and LFV) must be realized at some energy scale supersymmetrically . We review in this chapter avenues for neutrino mass generation in the framework of supersymmetry.

Unless neutrinos have "boring" Dirac masses $y_{\nu_{ab}} \langle H_2^0 \rangle L_a N_b$, which only imply super-light sterile right-handed neutrinos N_a and an extension of the usual fermion mass hierarchy problem, then the neutrinos have a $\Delta L = 2$ Majorana mass and there must be some source of total lepton number violation (LNV), $\Delta L \neq 0$. We will review both scenarios, R_P-conserving $\Delta L = 2$ terms and R_P-violating $\Delta L = 1$ terms.

We will also illustrate the "rewards" of the different scenarios: Supersymmetry provides somewhat delicate relations between neutrino physics and other arenas (such as LFV and $\tan\beta$-dependent observables in general) in the $\Delta L = 2$ case of a heavy right-handed neutrino, and many new channels for sparticle production and decay arise in the $\Delta L = 1$ R_P-violating case. Therefore, neutrino physics in supersymmetry is not independent but rather linked to other "new physics" observables, and it may be probed via various avenues.

11.1 $\Delta L = 2$ Theories

It is straightforward to "supersymmetrize" old ideas of a heavy right-handed (SM-singlet) neutrino with a large $\Delta L = 2$ Majorana mass M_R. The right-handed neutrino mixes with the SM left-handed neutrinos via the usual $\mathcal{O}(M_{\text{Weak}})$ Dirac mass term,

$$M_{(\nu_L \, \nu_R)} = \begin{pmatrix} & M_{\text{Weak}} \\ M_{\text{Weak}} & M_R \end{pmatrix}$$

leading to $m_\nu \sim M_{\text{Weak}}^2 / M_R$: The *see-saw* mechanism [149]. (Note that the mass and interaction eigenstates are nearly identical.) This mechanism does not require supersymmetry, but is trivially embedded in a supersymmetric framework by extending the superpotential (above the heavy neutrino scale M_R) $W \to W + y_{\nu_{ab}} H_2 L_a N_b + M_{R_{ab}} N_a N_b$ (and N_a here is the heavy right-handed neutrino superfield of generation $a = 1, 2, 3$).

Unified models (Chap. 6) typically imply further that the Dirac τ-neutrino mass is of the order of the top mass (top - neutrino unification), $m_{\nu_\tau} \sim m_t^2/M_R$. A natural choice for the scale M_R is the unification scale, and indeed simple fits to the data favor [5, 6] $M_R \sim 10^{13-15}$ GeV. Model building along these lines was a subject of intense activity following the observation of neutrino oscillations. For an example of a (unified) see-saw model, see Ref. [150], where a double see-saw mechanism $M_R \sim M_U^2/M_P$ and $m_\nu \sim M_W^2/M_R \sim M_W^2 M_P/M_U^2$ was proposed, with M_P is replaced by a lower "string scale". Another possibility [103] is that the scale M_R is related to the scale of spontaneous supersymmetry breaking in the hidden sector, in which case the Dirac mass may not be identical to m_t. "See-saw" models of neutrino masses were reviewed recently in Ref. [151].

Top – neutrino unification $y_t(M_U) = y_{\nu_\tau}(M_U)$ is natural in many unified models (but not $SU(5)$ where N is a singlet) in which the top and the right-handed neutrino are embedded in the same representation. We note that it excludes (independent of all other considerations such as the Higgs mass) the $\tan \beta \sim 1$ solutions to $b - \tau$ unification (Fig. 6.2). This double Yukawa unification constrains $\tan \beta$ from below due to cancellation of large top-Yukawa effects by the large neutrino-Yukawa effects [152]: In these models wave-function renormalization of the b by a top loop is balanced (above the heavy-neutrino decoupling scale) by the wave-function renormalization of the τ by a neutrino loop with $y_\nu \simeq y_t$. Eq. (6.8) is modified accordingly (in the decades between the unification M_U and neutrino M_R scales) to read

$$\frac{d}{d \ln \Lambda} \left(\frac{y_d}{y_l} \right) = \frac{1}{16\pi^2} \left(\frac{y_d}{y_l} \right) \times$$
$$\left\{ (y_u^2 - y_\nu^2) + 3(y_d^2 - y_l^2) - \frac{16}{3} g_3^2 - \frac{4}{3} g_1^2 \right\}. \quad (11.1)$$

In particular, it predicts $\tan \beta \gg 1$ and therefore that the light-Higgs mass saturates its upper bound of 130 GeV or so, consistent with experimental bounds. It also constrains other $\tan \beta$ dependent observables such as the anomalous muon magnetic moment (Sect. 9.2) where large values of $\tan \beta$ may lead to upper bounds on sparticle masses.

If indeed the right-handed neutrino is present, in supersymmetry there is also a right-handed sneutrino. Since the heavy right-handed neutrino decouples in a global supersymmetric regime, the sneutrino is also heavy and decouples. Nevertheless, it still has a "small" SSB mass. (This is not as straightforward in the case of low-energy gauge mediation.) It does not affect the decoupling yet it renormalizes other SSB terms of the light fields: We have already seen above that if the right-handed neutrino superfield couples with a large Yukawa coupling it can affect the renormalization of the superpotential (e.g. $b - \tau$ unification). Its renormalization of the SSB parameters is of similar origins and importance. It occurs in the decades in between the SSB mediation scale and the right-handed neutrino decoupling scale. For example, consider (i) RSB and the $m_{H_2}^2$ parameter which is now driven negative by both top and neutrino terms, or (ii) the slepton spectrum $m_{\tilde{l}}^2$.

A right-handed neutrino is therefore expected to leave its imprints in the weak-scale parameters such as $m_{\tilde{l}}^2$. In particular, since $y_{\nu_a} \neq y_{\nu_b}$ for $a \neq b$ then the affect are not flavor blind and one expects that it would lead to LFV effects. Particularly interesting is the disturbance of slepton (mass) universality (when assumed) and mass relations. (Relevant references were included in Ref. [127].) This in turn offers an interesting complementarity between neutrino physics, the slepton spectrum, and lepton flavor violation experiments (e.g. $\mu \to e\gamma$ discussed in Sect. 9.2 on the one hand and slepton oscillations [148] in future collider experiments on the other hand).

11.2 $\Delta L = 1$ Theories

The relation between neutrino physics and supersymmetry is even more fundamental and extensive if the neutrino mass originates from R-partiy violation [68, 63, 153, 154]. In this case the origin of the mass is at the weak scale and no right-handed neutrinos are necessary.

As we argued earlier, supersymmetry essentially encodes lepton - Higgs duality, which is typically removed by hand when imposing R-parity. Generically, the theory contains explicit $\Delta L = 1$ breaking via superpotential operators, SSB operators, and as a consequence, L could also be broken spontaneously by sneutrino vev's. One needs to apply $\Delta L = 1$ operation twice in order to induce a Majorana neutrino mass. In general, such models can admit [68] a one-loop radiative mass ($\propto \lambda'^2$) and a tree-level mass induced by neutrino-neutralino tree-level mixing $\propto \langle \tilde{\nu} \rangle \mu_L$ (Ex. 5.17). (Our notation follows eq. (5.15).)

The tree-level mass is intriguing: Only a supersymmetric neutrino mass $\mu_L L H_2$ is allowed in the superpotential (aside from the usual Higgs mass), while the neutrino is the only SM fermion which is left massless by the superpotential Yukawa terms. Hence, it offers an interesting complementarity between Yukawa and mass terms in the superpotential, and an elegant and economic realization of R-parity violation. It can also be shown to be related to the μ-problem. (For a discussion see Ref. [153, 155, 156] and references therein.) In the basis in which the electroweak breaking vev is only in the Higgs fields $\langle \tilde{\nu} \rangle = 0$ one has for the tree-level Majorana mass $m_\nu \simeq (\mu_L^2 M_W^2 \cos^2 \beta)/(2\mu^2 M_2)$. Note that only a single neutrino mass can be generated from tree-level neutralino-neutrino mixing. (This is because a chiral-like $SU(4)$ symmetry of the mass matrix is broken at tree-level at most to $SU(2)$, guaranteeing two massless states.)

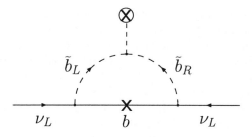

Fig. 11.1. A one-loop contribution to the Majorana neutrino mass arising from the $\Delta L = 1$ λ'_{i33} Yukawa operator.

In the remaining of this chapter, however, we focus on the radiative Majorana neutrino mass illustrated in Fig. 11.1. One obtains

$$\frac{m_\nu}{\text{MeV}} \sim \lambda'^2 \left(\frac{300\,\text{GeV}}{m_{\tilde{b}}} \right) \left(\frac{m^2_{LR}}{m_b\, m_{\tilde{b}}} \right), \tag{11.2}$$

where a b-quark and $\tilde{b}$-squark are assumed to circulate in the loop (so that $\lambda' = \lambda'_{i33}$), and m^2_{LR} is the $\tilde{b}$ left-right mixing squared mass. (The neutrino mass may vanish in the limit of a continuous $U(1)_R$ symmetry, which suppresses A and μ terms and corresponds in our case to $m^2_{LR} \ll m_b\, m_{\tilde{b}}$.) Imposing laboratory limits on the ν_e mass one can, for example, derive severe constraints on λ'_{133} [157], but λ'_{333} (i.e., $m_\nu \to m_{\nu_\tau}$) could still be $\mathcal{O}(1)$ [158]. (Note that a heavy $\mathcal{O}(\text{MeV})$ τ-neutrino has not been ruled out [7] but it cannot be stable on a cosmological time scales.) One has to consider the full λ' matrices in order to obtain neutrino mixing, but clearly, the complexity of the parameter space admits many possible scenarios for mixing[1].

R-partiy violation can then explain the neutrino spectrum as an electroweak (rather than GUT) effect. This comes at a price and with a reward. The price is that the sparticles are not stable and the LSP is not a dark matter candidate (but some other sector may still provide a dark matter candidate whose mass is of the order of the gravitino mass). The obvious reward is the many new channels are available at colliders to produce squarks and sleptons [63]. Consequently, the corresponding cross-section, branching-ratio, and asymmetry measurements may provide information on neutrino physics. We will conclude our discussion of LNV with an explicit example.

---

[1] Note that in the case that R-parity violation originating from a μ-term $W_{\Delta L=1} = \mu_L L H_2$ then the λ and λ' couplings appear by rotations of the usual Yukawa couplings once leptons and Higgs bosons are appropriately defined at low energies, e.g. H_1 is customarily defined along the relevant EWSB *vev*. Flavor mixing arises in this case as well at the loop level only.

11.3 Neutrinos vs. Collider Physics

As mentioned above, rich collider phenomena arise in the $\Delta L = 2$ case (e.g. slepton oscillations) and even more so in the case of $\Delta L = 1$ operators. In this section we focus on the latter. Perhaps the most relevant example, which can be probed in the current Tevatron run, is given by new exotic top decays. Application of $SU(2)$ and supersymmetry rotations to the vertex $\lambda'\nu b\tilde{b}$ assumed in (11.2), gives (the relevant vertex for) the decay channel $t_L \to b_R\tilde{\tau}_L$, assuming that it is kinematically allowed. Such decays were studied in Ref. [159, 158], and more recently in Ref. [160].

For $m_t = 175$ GeV one finds [158]

$$\frac{\Gamma_{t\to b\tilde{\tau}}}{\Gamma_{t\to bW}} = 1.12\,\lambda'^2_{333}\left[1 - \left(\frac{m_{\tilde{\tau}_L}}{175\text{GeV}}\right)^2\right]^2, \tag{11.3}$$

where explicit generation indices were introduced for clarity. The $\tilde{\tau}$-decay modes are highly model-dependent ·in the case of LNV. In particular, all superpartners typically decay in the collider and the typical large missing energy signature is replaced with multi-b and lepton signatures, which may be used for identification. (See, for example, Ref. [161].) If the sneutrino is the LSP then the three-body decays $\tilde{\tau} \to \tilde{\nu}_\tau f\bar{f}'$ and $\tilde{\tau} \to Wb\bar{b}$ are sufficiently phase space-suppressed (recall that $SU(2)$ invariance requires $m_{\tilde{\tau}_L} \simeq m_{\tilde{\nu}_\tau}$) so that the dominant decay mode is either $\tilde{\tau} \to \bar{c}b$ or $\tilde{\tau} \to e\bar{\nu}_e$.

In order to study the constraints on λ'_{333}, which are otherwise weak, let us assume that the $\bar{c}b$ mode is dominant. This new decay mode alters the number of $t\bar{t}$ events expected in each of the channels which characterize top decays (i.e., $tt \to$ jet jet, lepton lepton, lepton jet, where jet refers to hadronic activity) both through an enhancement of the percentage of hadronic decays and through the increased probability of b-tagging events given the assumption of b-rich $\tilde{\tau}$ decays. For each (final-product) channel, one can constraint λ'_{333} (as a function of the stau mass) from the number of events expected in the presence of $\tilde{\tau}$ decays relative to the number expected in the SM. (Currently, limited data on top decays is available and the constraints are still quite weak.) A more promising approach is to examine kinematic parameters in $t\bar{t}$ events, e.g., the reconstructed W mass in lepton + jets events with a second loosely tagged b [162]. (The two untagged jets define M_W.) At the time of the writing of this manuscript, with just 10 events, this gives $\lambda'_{333} \lesssim 0.4\,(1.0)$ for $m_{\tilde{\tau}_L} = 100\,(150)$ GeV [158]. Such kinematic analysis may therefore provide strong constraints on LNV couplings in the future.

Similarly to top decays, a $\tilde{\tau}$ may be produced in a larger mass range if it is radiated of a t or b quark, e.g. $gg, qq \to t(t \to b\tilde{\tau})$. (These channels compete with resonant (s-channel) production $qq \to \tilde{\tau}$ [163].) This was studied recently in Ref. [164]. The inclusive production cross sections $p\bar{p} \to t\tilde{\tau}_L X$, $pp \to t\tilde{\tau}_L X$ are obtained by combining the production cross sections arising from the $2 \to 2$ elementary process $gb \to t\tilde{\tau}_L$ to those induced by $2 \to 3$ partonic processes gg, $qq \to tb\tilde{\tau}_L$. (They may be independently measured if relatively complicated final states could be distinguished.) The corresponding inclusive cross sections are obtained by convolution of the hard-scattering cross section of quark- and gluon-initiated processes with the quark and gluon distribution functions in p and $\bar{p}$, and can be found in Ref. [164] where they are evaluated for the Tevatron and LHC parameters: See Figs. 11.2 and 11.3, respectively.

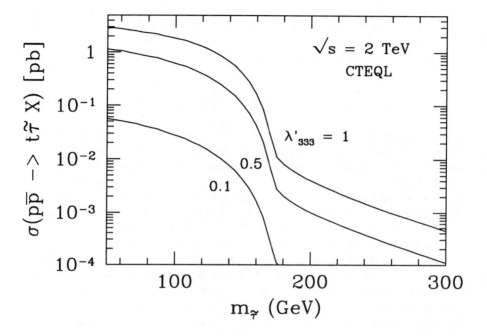

Fig. 11.2. The leading-order production cross section $\sigma(p\bar{p} \to t\tilde{\tau}X)$, for $\sqrt{s} = 2\,\text{TeV}$, as a function of the $\tilde{\tau}$ mass, is shown for different values of λ'_{333}. Renormalization and factorization scales are fixed as $\mu_R = \mu_f = m_t + m_{\tilde{\tau}}$. Taken from Ref. [164].

The large cross section obtained in the case of $\sqrt{s} = 14\,\text{TeV}$ implies that at the LHC, with a luminosity of $100\,\text{fb}^{-1}$ per year, light $\tilde{\tau}$'s may be produced in abundance even for couplings as small as 0.01, whereas for large couplings, they may be produced up to masses of $\mathcal{O}(1)\,\text{TeV}$. (Of course, background studies are needed before any final conclusions are drawn.) The Tevatron amplitude, on the other hand, is still dominated by the on-shell top decays discussed above, and no significant improvement in sensitivity is achieved.

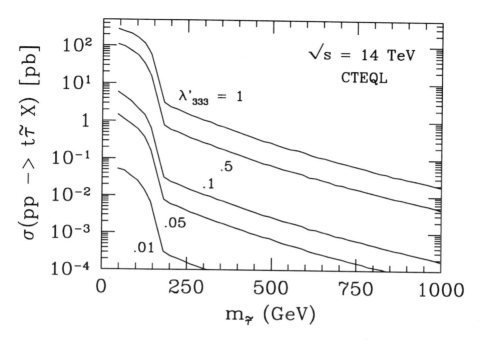

Fig. 11.3. The leading-order production cross section $\sigma(pp \to t\tilde{\tau}X)$, for $\sqrt{s} =$ 14 TeV, as a function of the $\tilde{\tau}$ mass is shown for different values of λ'_{333}. Renormalization and factorization scales are fixed as $\mu_R = \mu_f = m_t + m_{\tilde{\tau}}$. Taken from Ref. [164].

12. Vacuum Stability

In general, the global minimum of the scalar potential need not be the EWSB minimum, and it is far from clear that it is justified to consider only the Higgs potential when determining the vacuum[1]: The vacuum could break, in principle, color and/or charge.

Consider, as an example [165], the superpotential

$$W = y_t U_{L_3} U_3 H_2^0 + \mu H_1^0 H_2^0, \tag{12.1}$$

and we have performed an $SU(2)$ rotation so that H_2^+ has no vacuum expectation value. (The inclusion of H_1^- is therefore immaterial: We assume below that $m_1^2 > 0$.) In practice, eq. (12.1) includes the relevant terms in the superpotential when searching for the global minimum of the scalar potential in the large y_t limit, which will suffice here as an example. Using standard techniques we arrive at the corresponding scalar potential:

$$y_t^2 V = M^2 - \Gamma + \lambda^2, \tag{12.2}$$

where the bilinear, trilinear, and quartic terms are

$$M^2 = m_1^2 H_1^{0^2} + m_2^2 H_2^{0^2} - 2m_3^2 H_1^0 H_2^0 + m_{\tilde{t}_L}^2 \tilde{t}_L^2 + m_{\tilde{t}_R}^2 \tilde{t}_R^2, \tag{12.3}$$

$$\Gamma = \left| 2 \left(|A_t| H_2^0 + s|\mu| H_1^0 \right) \tilde{t}_L \tilde{t}_R \right|, \tag{12.4}$$

and

$$\lambda^2 = (\tilde{t}_L^2 + \tilde{t}_R^2) H_2^{0^2} + \tilde{t}_L^2 \tilde{t}_R^2 + \frac{1}{y_t^2} (\frac{\pi}{2} \times \text{``}D - \text{terms''}), \tag{12.5}$$

respectively. Here, $\tilde{t}_L = \tilde{U}_{L_3}$, $\tilde{t}_R = \tilde{U}_3 = \tilde{U}_{R_3}$. All fields were scaled $\phi \to \phi/y_t$ and are taken to be real and positive (our phase choice for the fields, which fixed $m_3^2 > 0$ and $\Gamma > 0$, i.e., maximized the negative contributions to V) and all parameters are real. $m_{1,2}^2 = m_{H_{1,2}}^2 + \mu^2$, $s = \mu A_t / |\mu A_t|$, and the expression for the "D-terms" is

$$4\alpha' \left[-\frac{H_1^{0^2}}{2} + \frac{H_2^{0^2}}{2} + \frac{\tilde{t}_L^2}{6} - \frac{2\tilde{t}_R^2}{3} \right]^2$$

---

[1] Sneutrino *vev's* discussed in the previous chapter are a (trivial) example.

$$+\alpha_2 \left[H_1^{0^2} - H_2^{0^2} + \tilde{t}_L^2 \right]^2$$
$$+\frac{4}{3}\alpha_3 \left[\tilde{t}_L^2 - \tilde{t}_R^2 \right]^2. \tag{12.6}$$

We can parameterize the four fields in terms of an overall scale ϕ and three angles $0 \le \alpha, \beta, \gamma \le \frac{\pi}{2}$: $H_1^0 = \phi \sin\alpha \cos\beta$, $H_2^0 = \phi \sin\alpha \sin\beta$, $\tilde{t}_R = \phi \cos\alpha \cos\gamma$, $\tilde{t}_L = \phi \cos\alpha \sin\gamma$, and redefine

$$y_t^2 V(\phi) = M^2(\alpha, \beta, \gamma)\phi^2 - \Gamma(\alpha, \beta, \gamma)\phi^3 + \lambda^2(\alpha, \beta, \gamma)\phi^4. \tag{12.7}$$

Then, for fixed angles, $V(\phi)$ will have a minimum for $\phi \ne 0$ provided the condition $32M^2\lambda^2 < 9\Gamma^2$ is satisfied. In that case,

$$\phi_{\min} = \frac{3}{8}\frac{\Gamma}{\lambda^2} \left[1 + \left(1 - \frac{32M^2\lambda^2}{9\Gamma^2} \right)^{\frac{1}{2}} \right] \ge 0, \tag{12.8}$$

and

$$y_t^2 V_{\min} = -\frac{1}{2}\phi_{\min}^2 \left(\frac{\Gamma}{2}\phi_{\min} - M^2 \right). \tag{12.9}$$

The SM minimum corresponds to $\alpha = \frac{\pi}{2}$ and $\beta = \beta^0$ (γ is irrelevant and $\tan\beta^0 = \nu_1/\nu_2$ is the angle (5.1) used to fix μ, m_3^2, as well as the Yukawa couplings). It is easy to convince oneself that in that limit the 4×4 second-derivative matrix is 2×2 block diagonal (otherwise baryon number is violated). Thus, it is sufficient to confirm that the four physical eigenvalues are positive to ensure that it is a minimum. If these conditions are satisfied then the SM is at least a local (negative-energy) minimum, and one has $\Gamma_{\rm SM} = 0$, $M_{\rm SM}^2 < 0$, and eqs. (12.8) and (12.9) reduce to the usual results $\phi_{\min}^{\rm SM} = \sqrt{-M_{\rm SM}^2/2\lambda_{\rm SM}^2}$, $y_t^2 V_{\min}^{\rm SM} = -M_{\rm SM}^4/4\lambda_{\rm SM}^2$.

Let us now consider the possibility of additional color and/or charge breaking (CCB) minima defined by $\cos\alpha \ne 0$. Assuming $m_{\tilde{t}_{L,R}}^2 > 0$, this requires $\Gamma \ne 0$, which we assume hereafter. From (12.7)–(12.9) it is easy to classify the possible CCB minima for definite α, β, γ. One finds that for $\Gamma^2 \le 32\lambda^2 M^2/9$ there is no CCB minimum, while for $32\lambda^2 M^2/9 < \Gamma^2 \le 4\lambda^2 M^2$ the CCB minimum exists but has $V_{\min}^{\rm CCB} > 0 > V_{\min}^{\rm SM}$, which is presumably safe. For $4\lambda^2 M^2 < \Gamma^2$ (including the more rare case $M^2 < 0$, that must fall in this category) there is a negative-value CCB minimum, which may however be either local (presumably safe), i.e., $V_{\min}^{\rm CCB} > V_{\min}^{\rm SM}$, or global (probably unacceptable), i.e., $V_{\min}^{\rm CCB} < V_{\min}^{\rm SM}$. Here, a sufficient (but not necessary) condition for an acceptable model is

$$\Gamma^2 \le 4\lambda^2 M^2. \tag{12.10}$$

In principle, the above discussion holds for any number of fields (i.e., any number of angles), only the explicit expressions for M^2, Γ, and λ^2 are more complicated. If constraint (12.10) holds for every choice of α, β, and γ ($\cos\alpha \ne 0$) then there is no negative-valued color and/or charge breaking

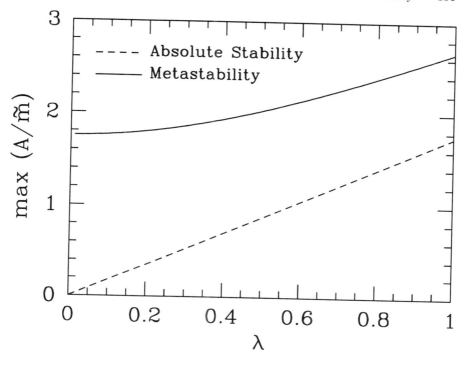

Fig. 12.1. Maximal value of $A/\widetilde{m}$ for absolute stability and metastability of tree-level scalar potential as a function of the square root of the quartic coupling, λ. Absolute stability refers to the absence of global charge-breaking minima, while metastability refers to a lifetime of the charge preserving vacuum greater than the age of the Universe, corresponding to a bounce action $S > 400$. Taken from Ref. [61].

minimum, global (GCCB) or local. In the special case $M^2 < 0$ the constraint cannot be satisfied and there will be a negative-energy local CCB minimum.

Eq. (12.10) is illustrated (for $M^2 = 3\widetilde{m}^2$ and $\Gamma = 2A$, as is appropriate for $H_2^0 = \tilde{t}_L = \tilde{t}_R$ and $H_1^0 = 0$; see below) by the lower curve in Fig. 12.1 (taken from Ref. [61]). The Higgs mass term m_i^2 receives a contribution from the μ-term. Hence, $\widetilde{m}^2 \gg m_{\tilde{t}_L, \tilde{t}_R}^2$ is possible in principle with a large Dirac mass, $\mu^2 \gg m_{\tilde{t}_L, \tilde{t}_R}^2$. However, minimal tuning of electroweak symmetry breaking (see Chap. 8) usually implies $|m_i^2| \sim M_Z^2 \lesssim m_{\tilde{t}_L, \tilde{t}_R}^2$. In what follows $\widetilde{m} \sim m_{\tilde{t}_L, \tilde{t}_R}$ is implicitly assumed.

If eq. (12.10) does not hold, further investigation is needed to determine whether the minimum is global or local. As long as m_2^2 is the only possible negative mass-squared parameter then eq. (12.10) is satisfied for $\sin \beta = 0$, i.e., there is no negative-valued CCB minimum for $H_2^0 = 0$. We will therefore restrict our attention to the case $H_2^0 \neq 0$, in which case it is convenient to reparameterize $\phi = H_2^0$ and rescale all fields by an additional factor of $1/H_2^0$, which simplifies the discussion.

One can derive analytic constraints from (12.10), which are typically relevant only for specific regions of the parameter space. The negative contribution of the trilinear terms to the total potential is maximized in a direction along which all scalar fields **that appear in the trilinear term** have equal expectation values (and all other fields vanish), so that the D-terms $\propto 1/y_t^2$ may vanish[2]. For example, if we fix $H_1^0 = 0$, $\tilde{t}_L = \tilde{t}_R = 1$ (all in H_2^0 units), then (12.10) gives the well-known result [96]

$$A_t^2 \leq 3(m_{\tilde{t}_L}^2 + m_{\tilde{t}_R}^2 + m_{H_2}^2 + \mu^2). \tag{12.11}$$

The equal-field direction is not a relevant requirement, however, for $y_t \approx 1$ (*i.e.*, there may exist a deeper minimum with non-vanishing D−terms). It is also obvious that eq. (12.11) is trivial in the limit $|\mu| \to \infty$. A more useful constraint in this case is [166] [taking $H_1^0 \approx 1$, $\tilde{t}_L \approx \tilde{t}_R \equiv t \ll 1$ and using (12.10)]

$$(|A_t| + s|\mu|)^2 \leq 2(m_{\tilde{t}_L}^2 + m_{\tilde{t}_R}^2). \tag{12.12}$$

Note that if the order t^2 corrections to (12.12) are not negligible then V is less negative, which motivated our choice $t \ll 1$, *i.e.*, the more dangerous direction. Other constraints are derived in different parts of the parameter space, particularly when the whole superpotential is considered.

Lastly, there are two caveats in our discussion above. The first is that we used in this section the (one-loop improved) tree-level potential. Calculations should therefore be performed at the $\tilde{t}$-scale to eliminate large loop corrections. Secondly, a GCCB minimum may be "safe" if separated from the local standard-model minimum by a tunneling time greater than the age of the universe.

Cosmological selection of the color and charge preserving vacuum at the origin is natural, since this is a point of enhanced symmetry and always a local minimum of the free energy at high enough temperature. It is sufficient therefore to ensure a lifetime greater than the present age of the universe. This corresponds to a bounce action out of the metastable vacuum of $S \gtrsim 400$ [167]. The bounce action for the potential under discussion may be calculated numerically [167, 168] and it interpolates between the thin-wall limit, $\lambda^2 \tilde{m}^2/A^2 \to 1/3^-$, in which the two vacua are nearly degenerate

$$S_{\text{thin}} \simeq \frac{9\pi^2}{2} \left(\frac{\tilde{m}^2}{A^2} \right) \left(1 - 3\lambda^2 \frac{\tilde{m}^2}{A^2} \right)^{-3},$$

and the thick-wall limit, $|\lambda^2 \tilde{m}^2/A^2| \ll 1$, in which the quartic term is not important

---

[2] Note that unlike the case of A-terms discussed here, the hypercharge D-term does not vanish along an equal-field direction $H_1^0 = \tilde{t}_L = \tilde{t}_R$ ($H_2^0 = 0$) corresponding to the operator $CH_1^{0*}\tilde{t}_L\tilde{t}_R$. Therefore, models with C-terms are generically more stable.

$$S_{\text{thick}} \simeq 1225 \left(\frac{\widetilde{m}^2}{A^2} \right).$$

(Here, we again assume for simplicity $M^2 = 3\widetilde{m}^2$, $\Gamma = 2A$, and the equal field direction of eq. (12.11).) The maximum value of $A/\widetilde{m}$ for which $S > 400$ is shown in fig. 12.1 as a function of the square root of the quartic coupling, λ (upper curve). (The figure is taken from Ref. [61] and the empirical fit to the numerically calculated bounce action given in Ref. [168] was employed.) The maximum allowed $A/\widetilde{m}$ continues as a smooth monotonically decreasing function for $\lambda^2 < 0$ (as might be the case if λ^2 arises along certain direction in field space only at one loop), even though the renormalizable potential is unbounded from below in this case. For $|\lambda^2| \ll 1$ the weak requirement of metastability on a cosmological time scale is met for $A/\widetilde{m} \lesssim 1.75$ ($|\Gamma/\sqrt{M^2}| \lesssim 2$).

Recommended readings on the subject of vacuum stability include Refs. [165, 167, 169].

13. Supersymmetry Breaking Operators

While supersymmetric extensions of the Standard Model can be fully described in terms of explicitly broken global supersymmetry, this description is only effective. Once related to spontaneous breaking in a more fundamental theory, the effective parameters translate to functions of two distinct scales, the scale of spontaneous supersymmetry breaking and the scale of its mediation to the standard-model fields. Here, the scale dependence will be again written explicitly, and the full spectrum of supersymmetry breaking operators which emerges will be explored, expanding on the discussion in Chap. 10.

Scale-dependent operators can play an important role in determining the phenomenology. For example, theories with low-energy supersymmetry breaking, such as gauge mediation, may correspond to a scalar potential which is quite different than in theories with high-energy supersymmetry breaking, such as gravity mediation. As a concrete example, the Higgs mass prediction (Chap. 8) will be discussed in some detail and its upper bound will be shown to be sensitive to the supersymmetry breaking scale.

Spontaneous supersymmetry breaking is most conveniently parameterized in terms of a spurion field $X = \theta^2 F_X$ with a non vanishing F-vev, which, as discussed previously, is an order parameter of (global) supersymmetry breaking. All SSB parameters can be written as non-renormalizable operators which couple the spurion X to the SM superfields. The operators are suppressed by the scale of the mediation of supersymmetry breaking from the (hidden) sector, parameterized by the spurion, to the SM (observable) sector. The coefficients of the various operators are dictated by the nature of the interaction between the sectors (the agent) and by the loop-order at which it occurs. In the following, we will explicitly write all operators which couple the two sectors. This will allow us to consider a general form of the supersymmetry breaking potential, which still resolves the hierarchy problem. While we will again identify the operators which correspond to the SSB parameters, our focus will be on those operators which are not included in the minimal form of the potential (5.13).

Even though the SSB parameters appearing in (5.13) are functions of the relevant ultraviolet scales, their magnitude is uniquely determined by the assumption that supersymmetry stabilizes the weak scale against divergent quantum corrections, $m_{\mathrm{SSB}} \sim \mathcal{O}(M_{\mathrm{Weak}}) \sim \mathcal{O}(100)$ GeV etc. This places a constraint on the ratio of the supersymmetry breaking scale $\sqrt{F}$ and the scale of its mediation M such that $F/M \simeq M_{\text{Weak}}$, and it implies that no information can be extracted regarding either scale from the SSB parameters. (Specific ultraviolet relations between the various SSB parameters could still be studied using the renormalization group formalism.) The generalization of (5.13) introduces dimensionless hard supersymmetry breaking (HSB) parameters in the potential. Their magnitude depends on the various scales in a way which both provides useful information (unlike the SSB parameters) and does not destabilize the solution to the hierarchy problem.

13.1 Classification of Operators

It is convenient to contain, without loss of generality, all observable-hidden interactions in the non-holomorphic Kähler potential K, $\mathcal{L} = \int d^2\theta d^2\bar{\theta} K$ (which is not protected by non-renormalization theorems). We therefore turn to a general classification of K-operators, originally presented in Ref. [170]. (See also Refs. [171, 172].) We do not impose any global symmetries, which can obviously eliminate some of the operators, and we keep all operators which survive the superspace integration.

The derivation here differs from the one in Chap. 10 in that that (a) all operators are contained, for convenience, in the Kähler potential K (and superpotential W operators are rewritten as K operators $O_W \to X^\dagger O_W$, with the appropriate adjustment of power and size of M); and (b) the spurion is assumed to have only a F-vev, $X = \theta^2 F_X$. In Chap. 10 a non-singlet field $Z = \theta^2 F_Z$ was considered as well. For simplicity, here we consider only a singlet $X = \theta^2 F_X$. In general, operators $\sim (XX^\dagger)^n$ imply also $(ZZ^\dagger)^n$-operators, with the possible exception of Z charged under non-linear symmetries. Also, in order to not confuse gauginos λ with quartic couplings, the latter are denoted in this chapter by[1] κ.

The effective low-energy Kähler potential of a rigid $N = 1$ supersymmetry theory is given by

$$
\begin{aligned}
K = \ &K_0(X, X^\dagger) + K_0(\Phi, \Phi^\dagger) \\
&+ \frac{1}{M} K_1(X, X^\dagger, \Phi, \Phi^\dagger) \\
&+ \frac{1}{M^2} K_2(X, X^\dagger, \Phi, \Phi^\dagger) \\
&+ \frac{1}{M^3} K_3(X, X^\dagger, \Phi, \Phi^\dagger, D_\alpha, W_\alpha) \\
&+ \frac{1}{M^4} K_4(X, X^\dagger, \Phi, \Phi^\dagger, D_\alpha, W_\alpha) + \cdots
\end{aligned}
\tag{13.1}
$$

---

[1] The symbol κ is usually reserved for non-renormalizable couplings. Indeed, the quartic couplings here, though renormalizable themselves, arise from integrating out non-renormalizable hidden-observable interactions.

where, as before, X is the spurion and Φ are the chiral superfields of the low-energy theory. D_α is the covariant derivative with respect to the super-space chiral coordinate θ_α, and W_α is the gauge supermultiplet in its chiral representation (Ex. 4.7). Once a separation between supersymmetry breaking fields X and low-energy Φ fields is imposed (i.e., the assumption that super-symmetry is broken in a hidden sector), there is no tree-level renormalizable interaction between the two sets of fields, and their mixing can arise only at the non-renormalizable level $K_{l\geq1}$.

The superspace integration $\mathcal{L}_D = \int d^2\theta d^2\bar{\theta} K$ reduces K_1 and K_2 to the usual SSB terms, as well as the superpotential μ-parameter $W \sim \mu\Phi^2$, which were discussed in Chap. 10. It also contains Yukawa operators $W \sim y\Phi^3$ which can appear in the effective low-energy superpotential. These are sum-marized in Tables 13.1 and 13.2. (We did not include linear terms that may appear (see Sect. 10.4 for an example) in the case of a low-energy singlet superfield Φ_{singlet}.) Finally, the last term in Table 13.1 contains correlated quartic and Yukawa couplings. They are soft as they involve at most loga-rithmic divergences.

Integration over K_3 produces non-standard soft terms, for example, the C (or A') terms. These terms are soft unless the theory contains a pure singlet field, in which case they can induce a quadratically divergent linear term. They are summarized in Table 13.3. The integration over K_3 also generates contributions to the ("standard") A and gaugino-mass terms. These terms could arise at lower orders in $\sqrt{F}/M$ from integration over holomorphic func-tions (and in the case of A, also from K_1), e.g., eqs. (10.4) and (10.5). Note that in the presence of superpotential Yukawa couplings, a supersymmetry breaking Higgsino mass term $\tilde{\mu}\tilde{H}_1\tilde{H}_2$ can be rotated to a combination of μ and $C-$terms, and vice versa (Ex. 5.10).

Lastly, superspace integration over K_4 leads to dimensionless hard oper-ators. These are summarized in Table 13.4, and will occupy the remaining of this chapter. Table 13.4 also contains supersymmetry breaking gauge-Yukawa interactions $\sim \lambda\psi\phi^*$. This is equivalent to the HSB kinetic term for the gaug-inos which were discussed recently in Ref. [136]. (Note that HSB gaugino couplings [145, 173] as well as quartic [174] and other HSB couplings are also generated radiatively in the presence of SSB.)

Higher orders in $(1/M)$ can be safely neglected as supersymmetry and the superspace integration allow only a finite expansion in $\sqrt{F_X}/M$, that is $\mathcal{L} = f[F_X^n/M^l]$ with $n \leq 2$ and l is the index K_l in expansion eq. (13.1). Hence, terms with $l > 4$ are suppressed by at least $(x/M)^{l-4}$. We assume the limit $x \ll M$ for the supersymmetry preserving vev x so that all such operators can indeed be neglected and the expansion is rendered finite.

It is interesting to identify two phenomenologically interesting groups of terms in K, (i) those terms which can break the chiral symmetries and can generate Yukawa terms in the low-energy effective theory, and (ii) new sources for quartic interactions.

The relevant chiral symmetry breaking terms in tables 13.1 and 13.3 can be identified with A and C terms which couple the matter sfermions to the Higgs fields of electroweak symmetry breaking. The chiral symmetry breaking originates in this case in the scalar potential and propagates to the fermions at one loop [61]. More interestingly, a generic Kähler potential is also found to contain tree-level chiral Yukawa couplings. These include $\mathcal{O}(F_X/M^2)$ supersymmetry conserving and SSB couplings and $\mathcal{O}(F_X^2/M^4)$ HSB chiral symmetry breaking couplings, leading to new avenues for fermion mass generation [170].

Quartic couplings arise at $\mathcal{O}(F_X/M^2)$, from supersymmetry conserving operators in Table 13.1 (depending on F_Φ), and at $\mathcal{O}(F_X^2/M^4)$ from HSB couplings in Table 13.4. They can potentially alter the supersymmetry conserving nature of the quartic potential, in general (e.g., in (8.1),(12.5),(12.6)).

The relative importance of the HSB operators relates to a more fundamental question: What are the scales $\sqrt{F_X}$ and M? This will be addressed in an example below. However, before doing so we need to address a different question regarding the potentially destabilizing properties of the different HSB operators, which relates to the nature of the cut-off scale Λ_{UV}. Indeed, one has to confirm that a given theory is not destabilized when the hard operators are included, an issue which is interestingly model independent. In order to do so, consider the implications of the hardness of the operators contained in K_4. Yukawa and quartic couplings can destabilize the scalar potential by corrections Δm^2 to the mass terms of the order of

$$
\Delta m^2 \sim \begin{cases} \frac{\kappa}{16\pi^2}\Lambda_{\mathrm{UV}}^2 \sim \frac{1}{16\pi^2}\frac{F_X^2}{M^4}\Lambda_{\mathrm{UV}}^2 \sim \frac{1}{16\pi^2 c_m}m^2 \\[2ex] \frac{y^2}{16\pi^2}\Lambda_{\mathrm{UV}}^2 \sim \frac{1}{16\pi^2}\frac{F_X^4}{M^8}\Lambda_{\mathrm{UV}}^2 \sim \frac{1}{16\pi^2 c_m}m^2\frac{m^2}{M^2}, \end{cases} \tag{13.2}
$$

where we identified $\Lambda_{\mathrm{UV}} \simeq M$ and c_m is a dimensionless coefficient omitted in Table 13.1, $m^2/2 = c_m F_X^2/M^2$. The hard operators were substituted by the appropriate powers of F_X/M^2. Once M is identified as the cut-off scale above which the full supersymmetry is restored, then these terms are harmless as the contributions are bound from above by the tree-level scalar mass-squared parameters. In particular, the softness condition imposed on the supersymmetry breaking terms in Chap. 5 was sufficient but not necessary. (This observation extends to the case of non-standard soft operators such as $C \sim F_X^2/M^3$ in the presence of a singlet).

In fact, such hard divergent corrections are well known in supergravity with $\Lambda_{\mathrm{UV}} = M = M_P$, where they perturb any given set of tree-level boundary conditions for the SSB parameters (see Sect. 10.2 and references therein). Given the supersymmetry breaking scale in this case, $F \simeq M_{\mathrm{Weak}}M_P$, the Yukawa (and quartic) operators listed below are proportional in these theories to $(M_{\mathrm{Weak}}/M_P)^n$, $n = 1, 2$, and are often omitted. Nevertheless, they can shift any boundary conditions for the SSB by $\mathcal{O}(1 - 100\%)$ due to quadratically divergent one-loop corrections.

We conclude that, in general, quartic couplings and chiral Yukawa couplings appear once supersymmetry is broken, and if supersymmetry is broken at low energy then these couplings could be sizable yet harmless. We will explore possible implications of the HSB quartic couplings in the next section. In the next chapter we will touch upon some implications of HSB chiral Yukawa couplings.

Table 13.1. The soft supersymmetry breaking terms as operators contained in K_1 and K_2. $\Phi = \phi + \theta\psi + \theta^2 F$ is a low-energy superfield while X, $\langle F_X \rangle \neq 0$, parameterizes supersymmetry breaking. $F^\dagger = \partial W / \partial \Phi$. The infrared operators are obtained by superspace integration over the ultraviolet operators.

Ultraviolet K operator	Infrared $\mathcal{L}_D$ operator
$\frac{X}{M}\Phi\Phi^\dagger + h.c.$	$A\phi F^\dagger + h.c.$
$\frac{XX^\dagger}{M^2}\Phi\Phi^\dagger + h.c.$	$\frac{m^2}{2}\phi\phi^\dagger + h.c.$
$\frac{XX^\dagger}{M^2}\Phi\Phi + h.c.$	$B\phi\phi + h.c.$
$\frac{X^\dagger}{M^2}\Phi^2\Phi^\dagger + h.c.$	$\kappa\phi^\dagger\phi F + h.c.$ $y\phi^\dagger\psi\psi + h.c.$

Table 13.2. The effective renormalizable $N = 1$ superpotential W operators contained in K_1 and K_2, $\mathcal{L} = \int d^2\theta W$. Symbols are defined in Table 13.1. The infrared operators are obtained by superspace integration over the ultraviolet operators.

Ultraviolet K operator	Infrared W operator
$\frac{X^\dagger}{M}\Phi^2 + h.c.$	$\mu\Phi^2$
$\frac{X^\dagger}{M^2}\Phi^3 + h.c.$	$y\Phi^3$

13.2 The Higgs Mass vs. the Scale of Supersymmetry Breaking

As shown above, in general, HSB quartic couplings κ_{hard} arise in the scalar potential (from non-renormalizable operators in the Kähler potential, for ex-

Table 13.3. The non-standard or semi-hard supersymmetry breaking terms as operators contained in K_3. W_α is the $N = 1$ chiral representation of the gauge supermultiplet and λ is the respective gaugino. D_α is the covariant derivative with respect to the superspace coordinate θ_α. All other symbols are as in Table 13.1. The infrared operators are obtained by superspace integration over the ultraviolet operators.

Ultraviolet K operator	Infrared $\mathcal{L}_D$ operator
$\frac{XX^\dagger}{M^3}\Phi^3 + h.c.$	$A\phi^3 + h.c.$
$\frac{XX^\dagger}{M^3}\Phi^2\Phi^\dagger + h.c.$	$C\phi^2\phi^\dagger + h.c.$
$\frac{XX^\dagger}{M^3}D^\alpha\Phi D_\alpha\Phi + h.c.$	$\tilde{\mu}\psi\psi + h.c.$
$\frac{XX^\dagger}{M^3}D^\alpha\Phi W_\alpha + h.c.$	$M'_\lambda\psi\lambda + h.c.$
$\frac{XX^\dagger}{M^3}W^\alpha W_\alpha + h.c.$	$\frac{M_\lambda}{2}\lambda\lambda + h.c.$

Table 13.4. The dimensionless hard supersymmetry breaking terms as operators contained in K_4. Symbols are defined as in Table 13.1 and Table 13.3. The infrared operators are obtained by superspace integration over the ultraviolet operators.

Ultraviolet K operator	Infrared $\mathcal{L}_D$ operator
$\frac{XX^\dagger}{M^4}\Phi D^\alpha\Phi D_\alpha\Phi + h.c.$	$y\phi\psi\psi + h.c.$
$\frac{XX^\dagger}{M^4}\Phi^\dagger D^\alpha\Phi D_\alpha\Phi + h.c.$	$y\phi^\dagger\psi\psi + h.c.$
$\frac{XX^\dagger}{M^4}\Phi D^\alpha\Phi W_\alpha + h.c.$	$\bar{y}\phi\psi\lambda + h.c.$
$\frac{XX^\dagger}{M^4}\Phi^\dagger D^\alpha\Phi W_\alpha + h.c.$	$\bar{y}\phi^\dagger\psi\lambda + h.c.$
$\frac{XX^\dagger}{M^4}\Phi W^\alpha W_\alpha + h.c.$	$\bar{y}\phi\lambda\lambda + h.c.$
$\frac{XX^\dagger}{M^4}\Phi^\dagger W^\alpha W_\alpha + h.c.$	$\bar{y}\phi^\dagger\lambda\lambda + h.c.$
$\frac{XX^\dagger}{M^4}\Phi^2\Phi^{\dagger\,2} + h.c.$	$\kappa(\phi\phi^\dagger)^2 + h.c.$
$\frac{XX^\dagger}{M^4}\Phi^3\Phi^\dagger + h.c.$	$\kappa\phi^3\phi^\dagger + h.c.$

Table 13.5. Frameworks for estimating κ_{hard}. (Saturation of the lower bound on M is assumed.)

Framework	$\hat{\kappa}$	M	κ_{hard}
TLM $(n = 0)$	~ 1	$\gtrsim \widetilde{m}$	$(\widetilde{m}/M)^2 \sim 1$
NPGM $(n = 1/2)$	~ 1	$\gtrsim 4\pi\widetilde{m}$	$(4\pi\widetilde{m}/M)^2 \sim 1$
MGM $(n = 1)$	$\lesssim \frac{1}{16\pi^2}$	$\gtrsim 16\pi^2\widetilde{m}$	$(4\pi\widetilde{m}/M)^2 \sim \frac{1}{16\pi^2}$

ample). Assuming that the SSB parameters are characterized by a parameter $\widetilde{m} \sim 1\,\text{TeV}$ then

$$\kappa_{\text{hard}} = \hat{\kappa}\frac{F^2}{M^4} \simeq \hat{\kappa}(16\pi^2)^{2n}\left(\frac{\widetilde{m}}{M}\right)^2, \tag{13.3}$$

where M is a dynamically determined scale parameterizing the communication of supersymmetry breaking to the SM sector, which is distinct from the supersymmetry breaking scale $\sqrt{F} \simeq (4\pi)^n\sqrt{\widetilde{m}M}$. The exponent $2n$ is the loop order at which the mediation of supersymmetry breaking to the (quadratic) scalar potential occurs. (Non-perturbative dynamics may lead to different relations that can be described instead by an effective value of n.) The coupling $\hat{\kappa}$ is an unknown dimensionless coupling (for example, in the Kähler potential). As long as such quartic couplings are not arbitrary but are related to the source of the SSB parameters and are therefore described by (13.3), then they do not destabilize the scalar potential and do not introduce quadratic dependence on the ultraviolet cut-off scale, which is identified with M. This was demonstrated in the previous section. Stability of the scalar potential only constrains $\hat{\kappa} \lesssim \min\left((1/16\pi^2)^{2n-1}, 1\right)$ (though calculability and predictability are diminished).

The F- and D-term-induced quartic potential (e.g. in eq. (8.1)) gives for the (pure D-induced) tree-level Higgs coupling, $V = \kappa h^4$,

$$\kappa = \frac{g'^2 + g^2}{4}\cos^2 2\beta, \tag{13.4}$$

where we work in the decoupling limit in which one physical Higgs doublet H is sufficiently heavy and decouples from electroweak symmetry breaking while a second SM-like Higgs doublet is roughly given by $h \simeq H_1\cos\beta + H_2\sin\beta$. (This was explained in Chap. 8.) The HSB coupling corrects this relation. Given the strict tree-level upper bound that follows from (13.4), $m_{h^0}^2 \leq M_Z^2\cos^2 2\beta$, it is suggestive that HSB may not be only encoded in, but also measured via, the Higgs mass. We will explore this possibility, originally pointed out and studied in Ref. [78], in this section.

In the case that supergravity interactions mediate supersymmetry breaking from some hidden sector (where supersymmetry is broken spontaneously) to the SM sector, one has $M = M_P$. The corrections are therefore negligible whether the mediation occurs at tree level ($n = 0$) or loop level ($n \geq 1$) and can be ignored for most purposes. (For exceptions, see discussion and references of quantum effects in supergravity in Chap. 10 as well as Ref. [172].) In general, however, the scale of supersymmetry breaking is an arbitrary parameter and depends on the dynamics that mediate the SSB parameters. For example, in the case of $N = 2$ supersymmetry (see the next chapter) one expects $M \sim 1\,\text{TeV}$ [170]. Also, in models with extra large dimensions the fundamental M_{Planck} scale can be as low as a few TeV, leading again to $M \sim 1\,\text{TeV}$. (For example, see Ref. [28].) A "TeV-type" mediation scale implies a similar supersymmetry breaking scale and provides an unconventional possibility. (For a discussion, see Ref. [170].) If indeed $M \sim 1\,\text{TeV}$ then κ_{hard} given in (13.3) is $\mathcal{O}(1)$ (assuming tree-level mediation (TLM) and $\mathcal{O}(1)$ couplings $\hat{\kappa}$ in the Kähler potential). The effects on the Higgs mass must be considered in this case.

Though one may argue that TLM models represent a theoretical extreme, this is definitely a viable possibility. A more familiar and surprising example is given by the (low-energy) gauge mediation (GM) framework. (See Sect. 10.3). In GM, SM gauge loops communicate between the SM fields and some messenger sector(s), mediating the SSB potential. The Higgs sector and the related operators, however, are poorly understood in this framework [140] and therefore all allowed operators should be considered. In its minimal incarnation (MGM) $2n = 2$, and $M \sim 16\pi^2 \widetilde{m} \sim 100\,\text{TeV}$ parameterizes both the mediation and supersymmetry breaking scales. The constraint (13.2) corresponds to $\kappa_{\text{hard}} \sim \hat{\kappa} \lesssim 1/16\pi^2$ and the respective contribution to the Higgs mass could be comparable to the contribution of the supersymmetric coupling (13.4). A particularly interesting case is that of non-perturbative messenger dynamics (NPGM) in which case $n_{eff} = 1/2$, $M \sim 4\pi\widetilde{m} \sim 10\,\text{TeV}$ [138], and the constraint on $\hat{\kappa}$ is relaxed to 1. Now $\kappa_{\text{hard}} \lesssim 1$ terms could dominate the Higgs mass. The various frameworks are summarized in Table 13.5.

In order to address the β-dependence of the HSB contributions (which is different from that of all other terms) we recall the general two-Higgs-doublet model (2HDM). The Higgs quartic potential was already given in a general form in eq. 3.3. It can be written down as (e.g. see Ref. [174])

$$
\begin{aligned}
V_{\phi^4} = {} & \frac{1}{2}\kappa_1(H_1^\dagger H_1)^2 + \frac{1}{2}\kappa_2(H_2^\dagger H_2)^2 \\
& + \kappa_3(H_1^\dagger H_1)(H_2^\dagger H_2) + \kappa_4(H_1 H_2)(H_2^\dagger H_1^\dagger) \\
& + \left\{ \frac{1}{2}\kappa_5(H_1 H_2)^2 + [\kappa_6(H_1^\dagger H_1) \right. \\
& \left. + \kappa_7(H_2^\dagger H_2)]H_1 H_2 + h.c. \right\}.
\end{aligned}
\tag{13.5}
$$

In the decoupling limit it simply reduces to the SM with one "light" physical Higgs boson h^0, $m_{h^0}^2 = \kappa\nu^2$, $\kappa = c_\beta^4\kappa_1 + s_\beta^4\kappa_2 + 2s_\beta^2 c_\beta^2(\kappa_3 + \kappa_4 + \kappa_5) + 4c_\beta^3 s_\beta\kappa_6 +$

$4c_\beta s_\beta^3 \kappa_7$, where $s_\beta \equiv \sin\beta$ and $c_\beta \equiv \cos\beta$, and $\nu = \langle h^0 \rangle = 174$ GeV is the SM Higgs vev (normalized consistently with Chap. 8).

Allowing additional HSB quartic terms besides the usual gauge $(D\text{-})$terms and loop contributions, $\kappa_{1...7}$ can be written out explicitly as

$$\kappa_{1,2} = \frac{1}{2}(g'^2 + g^2) + \kappa_{\text{soft }1,2} + \kappa_{\text{hard }1,2}, \tag{13.6}$$

$$\kappa_3 = -\frac{1}{4}(g'^2 - g^2) + \kappa_{\text{soft }3} + \kappa_{\text{hard }3}, \tag{13.7}$$

$$\kappa_4 = -\frac{1}{2}g^2 + \kappa_{\text{soft }4} + \kappa_{\text{hard }4}, \tag{13.8}$$

$$\kappa_{5,6,7} = \kappa_{\text{soft }5,6,7} + \kappa_{\text{hard }5,6,7}, \tag{13.9}$$

where g' and g are again the SM hypercharge and $SU(2)$ gauge couplings, and $\kappa_{\text{soft }i}$ sums the loop effects due to soft supersymmetry breaking effects $\sim \ln\tilde{m}$, eqs. (8.10), (8.11). The effect of the HSB contributions $\kappa_{\text{hard }i}$ is estimated next.

While Ref. [78] explores the individual contribution of each of the κ_{hard} couplings, here we will assume, for simplicity, that $\kappa_{hard\,i} = \kappa_{\text{hard}}$ are all equal and positive. The squared Higgs mass $m_{h^0}^2$ reads in this case

$$m_{h^0}^2 = M_Z^2 \cos^2 2\beta + \delta m_{\text{loop}}^2 + (c_\beta + s_\beta)^4 \nu^2 \kappa_{\text{hard}}, \tag{13.10}$$

where $\delta m_{\text{loop}}^2 \lesssim M_Z^2$. (Note that no new particles or gauge interactions were introduced.)

Given the relation (13.10), one can evaluate the HSB contributions to the Higgs mass for an arbitrary M (and n). We define an effective scale $M_* \equiv (M/(4\pi)^{2n}\sqrt{\hat{\kappa}})(\text{TeV}/\tilde{m})$. The HSB contributions decouple for $M_* \gg \tilde{m}$, and the results reduce to the MSSM limit with only SSB (e.g. supergravity mediation). However, for smaller values of M_* the Higgs mass is dramatically enhanced. For $M{=}1$ TeV and TLM or $M{=}4\pi$ TeV and NPGM, both of which correspond to $M_* \simeq 1$ TeV, the Higgs mass could be as heavy as 475 GeV for $\tan\beta = 1.6$ and 290 GeV for $\tan\beta = 30$. This is to be compared with 104 GeV and 132 GeV [175], respectively, if HSB are either ignored or negligible. (A SM-like Higgs boson h^0 may be as heavy as 180 GeV in certain $U(1)'$ models [112] with only SSB.)

In the MGM case $\hat{\kappa} \lesssim 1/16\pi^2$ so that $M_* \sim 4\pi$ TeV (unlike the NPGM where $M_* \sim 1$ TeV). HSB effects are now more moderate but can increase the Higgs mass by 40 (10) GeV for $\tan\beta = 1.6$ (30) (in comparison to the case with only SSB.) Although the increase in the Higgs mass in this case is not as large as in the TLM and NPGM cases, it is of the same order of magnitude as, or larger than, the two-loop corrections due to SSB [175] (which are typically of the order of a few GeV), setting the uncertainty range on any such calculation. Also, it is more difficult now to set a model-independent lower bound on $\tan\beta$ based on the Higgs mass.

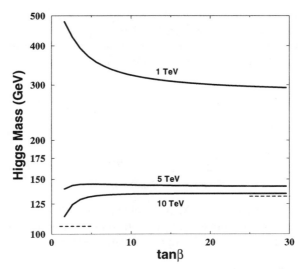

Fig. 13.1. The light Higgs boson mass (note the logarithmic scale) is shown as a function of $\tan\beta$ for $M_* = 1, 5, 10$ TeV (assuming equal HSB couplings). The upper bound when considering only SSB ($M_* \to \infty$) is indicated for comparison (dashed lines) for $\tan\beta = 1.6$ (left) and 30 (right). Taken from Ref. [78].

In Fig. 13.1, m_{h^0} dependence on $\tan\beta$ for fixed values of M_* is shown. The $\tan\beta$ dependence is from the tree-level mass and from the HSB corrections, while the loop corrections to $m_{h^0}^2$ are fixed, for simplicity, at $9200\,\text{GeV}^2$ [175]. The upper curve effectively corresponds to $\kappa_{\text{hard}} \simeq 1$. The HSB contribution dominates the Higgs mass and m_{h^0} decreases with increasing $\tan\beta$. As indicated above, m_{h^0} could be in the range of 300-500 GeV, dramatically departing from calculations which ignore HSB terms. The lower two curves illustrate the range[2] of the corrections in the MGM, where the tree-level and the HSB contributions compete. The $\cos 2\beta$ dependence of the tree-level term dominates the β-dependence of these two curves. Clearly, the Higgs mass could discriminate between the MGM and NPGM and help to better understand the origin of the supersymmetry breaking.

Following the Higgs boson discovery, it should be possible to extract information on the mediation scale M. In fact, some limits can already be extracted. Consider the upper bound on the Higgs mass derived from a fit to electroweak precision data: $m_h^0 < 215$ GeV at 95% confidence level [7]. (Such fits are valid in the decoupling limit discussed here.) A lower bound on the scale M in MGM could be obtained by rewriting eq. (13.10) as

---

[2] Given the many uncertainties, e.g. the messenger quantum numbers and multiplicity and $\sqrt{F}/M$ [137], we identify the MGM with a M_*-range which corresponds to a factor of two uncertainty in the hard coupling.

$$m_Z^2 \cos^2 2\beta + \delta m_{\text{loop}}^2 + (c_\beta + s_\beta)^4 \nu^2 \left(\frac{4\pi m_0}{M}\right)^2 \leq (215 \text{ GeV})^2, \qquad (13.11)$$

assuming equal κ_{hard}'s. For $\tan\beta = 1.6$, it gives $M \geq 31$ TeV while for $\tan\beta = 30$ the lower bound is $M \geq 19$ TeV. Once m_{h^0} is measured, more stringent bounds on M could be set.

Our discussion illustrates that the scale of the mediation of supersymmetry breaking explicitly appears in the prediction of the Higgs mass (and with a distinct β-dependence). In turn, it could lead in certain cases to a much heavier Higgs boson than usually anticipated in supersymmetric theories. It could also distinguish models, e.g. supergravity mediation from other low-energy mediation and weakly from strongly interacting messenger sectors. Given our ignorance of the (Kähler potential and) HSB terms, such effects can serve for setting the uncertainty on any Higgs mass calculations and can be used to qualitatively constrain the scale of mediation of supersymmetry breaking from the hidden to the SM sector.

14. $N = 2$ Supersymmetry

We conclude these notes by considering an extended supersymmetry framework, specifically, $N = 2$ supersymmetry, and by further entertaining the possibility of embedding the SM in such a framework.

The N=2 supersymmetry algebra has two spinorial generators Q^i_α, $i = 1, 2$, satisfying

$$\{Q^i_\alpha, \bar{Q}^j_{\dot\beta}\} = 2\sigma^\mu_{\alpha\dot\beta} P_\mu \delta^{ij}, \tag{14.1}$$

where σ^μ are, as usual, the Pauli matrices and P_μ is the momentum. The $N = 2$ theory can be described in the $N = 1$ formulation, as we will do here, as long as one imposes a non-Abelian R-symmetry, the exchange $SU(2)_R$ symmetry, on the $N = 1$ description. The supercharges Q^i_α form a doublet of the (exchange) $SU(2)_R$ R-symmetry, and the $N = 1$ description can be viewed as a rotation acting on it (while the Lagrangian preserves the full symmetry).

The lowest $N = 2$ spin representations, which are the relevant ones for embedding the SM, are the hypermultiplet and vector multiplet. Written in the familiar $N = 1$ language, the hypermultiplet is composed of two $N = 1$ chiral multiplets $X = (\phi_x, \psi_x)$ and $Y = (\phi_y, \psi_y)$, with Y occupying representation $\mathcal{R}$ of the gauge groups which are conjugate to that of X, $\mathcal{R}(X) = \mathcal{R}(Y^\dagger)$. Schematically, the hypermultiplet is described by a "diamond" plot

where the first, second and third rows correspond to helicity $-1/2$, 0, and $+1/2$ states, respectively. The vector multiplet contains a $N = 1$ vector multiplet $V = (V^\mu, \lambda)$, where λ is a gaugino, and a $N = 1$ chiral multiplet $\Phi_V = (\phi_V, \psi_V)$ in the adjoint representation of the gauge group (or a singlet in the Abelian case). Schematically, it is described by

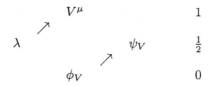

where the first, second and third rows correspond to helicity 0, 1/2, and 1 states, respectively. The $N = 1$ superfields are given by the two 45° sides of each diamond (indicated by arrows), with the gauge field arranging itself in its chiral representation W_α. The particle content is doubled in comparison to the $N = 1$ supersymmetry case and it is four times that of the SM: For each of the usual chiral fermions ψ_x and its complex-scalar partner ϕ_x, there are a conjugate mirror fermion ψ_y and complex scalar ϕ_y (so that the theory is vectorial). For each gauge boson and gaugino, there is a *mirror* gauge boson ϕ_V and a *mirror* gaugino ψ_V.

The $N = 0$ boson and fermion components of the hyper and vector-multiplet form $SU(2)_R$ representations. States with equal helicity form a $SU(2)_R$ doublet $(\phi_x, \phi_y^\dagger)$ and an anti-doublet (ψ_V, λ), while all other states are $SU(2)_R$ singlets. In fact, the full R-symmetry is $U(2)_R$ of which the exchange $SU(2)_R$ is a subgroup. There are additional $U(1)_R^{N=2}$, $U(1)_J^{N=2}$ subgroups such that the R-symmetry is either $SU(2)_R \times U(1)_R^{N=2}$, or in some cases only a reduced $U(1)_J^{N=2} \times U(1)_R^{N=2}$. The different superfields $X \sim \phi_x + \theta\psi_x$, etc. transform under the $U(1)$ symmetries with charges R and J given by

$$R(X) = r = -R(Y), \quad R(\Phi_V) = -2, \tag{14.2}$$

$$J(X) = -1 = J(Y), \quad J(\Phi_V) = 0, \tag{14.3}$$

and $R(W_\alpha) = J(W_\alpha) = -1$. The (manifest) supercoordinate θ has, as usual, charge $R(\theta) = J(\theta) = -1$.

The $SU(2)_R \times U(1)_R^{N=2}$ invariant $N = 2$ Lagrangian can be written in the $N = 1$ language as

$$\mathcal{L} = \int d^2\theta \left\{ \frac{1}{2g^2} W^\alpha W_\alpha + \sqrt{2} i g Y \Phi_V X + h.c. \right\}$$

$$+ \int d^2\theta d^2\bar\theta \left\{ \mathrm{Tr}(\Phi_V^\dagger e^{2gV} \Phi_V e^{-2gV} + X^\dagger e^{2gV} X + Y^\dagger e^{-2gV^T} Y) \right\}, \tag{14.4}$$

where here we wrote the gauge coupling g explicitly, the first (second) integral is the F- (D-) term, and we allow for non-Abelian gauge fields (not included in Chap. 4): $\Phi_V = \Phi_V^a T^a$ and $V = V^a T^a$, T^a being the respective generators. The second F-term is the superpotential. The only free coupling is the gauge coupling g: The coupling constant of the Yukawa term in the superpotential is fixed by the gauge coupling due to a global $SU(2)_R$. In particular, the $SU(2)_R$ symmetry forbids any chiral Yukawa terms so that fermion mass generation is linked to supersymmetry breaking (we return to this point below). Note that the $U(1)_R^{N=2}$ forbids any mass terms $W \sim \mu' XY$ (and the full R-symmetry forbids the usual $N = 1$ μ-term $W \sim \mu H_1 H_2$.) Unlike the $SU(2)_R$, $U(1)_R^{N=2}$ can survive supersymmetry breaking.

The $N = 2$ Lagrangian (14.4) also exhibits several discrete symmetries, which may or may not be broken in the broken supersymmetry regime. There is a trivial extension of the usual $N = 1$ R-parity Z_2 symmetry which does not distinguish the ordinary fields from their mirror partners:

$$\theta \to -\theta, \quad X_M \to -X_M, \quad Y_M \to -Y_M, \tag{14.5}$$

where all other supermultiplets are R_P-even and where the hypermultiplets have been divided into the odd matter multiplets (X_M, Y_M) and the even Higgs multiplets (X_H, Y_H). (Note that V is even but W_α is odd.) As in the $N = 1$ case, all the ordinary and mirror quarks, leptons and Higgs bosons are R_P-even, while the ordinary and mirror gauginos are R_P-odd. R_P is conveniently used to define the superpartners (or sparticles) as the R_P-odd particles. The LSP is stable if R_P remains unbroken. This was all discussed in Sect. 5.6.

A second parity, called mirror parity (M_P), distinguishes the mirror particles from their partners:

$$\theta \to \theta, \quad Y_M \to -Y_M, \quad Y_H \to -Y_H, \quad \Phi_V \to -\Phi_V, \tag{14.6}$$

and all other superfields (including W_α) are M_P-even. It is convenient to use mirror parity to define the mirror particles as the M_P-odd particles. (This definition should not be confused with other definitions of mirror particles used in the literature and which are based on a left-right group $SU(2)_L \times SU(2)_R$ or a mirror world which interacts only gravitationally with the SM world.) The lightest mirror parity odd particle (LMP) is also stable in a theory with unbroken mirror parity. However, if supersymmetry breaking does not preserve mirror parity, mixing between the ordinary matter and the mirror fields is allowed.

There is also a reflection (exchange) symmetry (which must be broken at low energies), the mirror exchange symmetry:

$$X \leftrightarrow Y, \quad \Phi_V \leftrightarrow \Phi_V^{\mathrm{T}}, \quad V \leftrightarrow -V^{\mathrm{T}}. \tag{14.7}$$

Like in the case of this continuous $SU(2)_R$, if the reflection symmetry remains exact after supersymmetry breaking then for each left-handed fermion there would be a degenerate right handed mirror fermion in the conjugate gauge representation, which is phenomenologically not acceptable.

For easy reference, we list in Table 14.1 the minimal particle content of the MN2SSM: The minimal $N = 2$ supersymmetric[1] extension of the SM. A

---

[1] Note that here we discuss the case of global $N = 2$ supersymmetry. In the local case there are also two gravitini and the relation between their mass and the SSB parameters is not obvious.

mirror partner Y (Φ_V) exists for every ordinary superfield X (V) of the $N = 1$ MSSM. One could eliminate one Higgs hypermultiplet and treat H_1 and H_2 as mirror partners. However, this could lead to the spontaneous breaking of mirror parity when the Higgs bosons acquire *vev's*, and as a result, to more complicated mixing and radiative structures than in a theory with two Higgs hypermultiplets.

Table 14.1. Hypermultiplets and vector multiplets in the MN2SSM. Our notation follows that of Table 1.1.

	X / V	Y / Φ_V
	$Q = (\widetilde{Q}, Q) = (3, 2)_{\frac{1}{6}}$	$Q' = (\widetilde{Q}', Q') = (\bar{3}, \bar{2})_{-\frac{1}{6}}$
Matter	$U = (\widetilde{U}, U) = (\bar{3}, 1)_{-\frac{2}{3}}$	$U' = (\widetilde{U}', U') = (3, 1)_{\frac{2}{3}}$
(hyper-)	$D = (\widetilde{D}, D) = (\bar{3}, 1)_{\frac{1}{3}}$	$D' = (\widetilde{D}', D') = (3, 1)_{-\frac{1}{3}}$
multiplets	$L = (\widetilde{L}, L) = (1, 2)_{-\frac{1}{2}}$	$L' = (\widetilde{L}', L') = (1, \bar{2})_{\frac{1}{2}}$
	$E = (\widetilde{E}, E) = (1, 1)_1$	$E' = (\widetilde{E}', E') = (1, 1)_{-1}$
Higgs (hyper -)	$H_1 = (H_1, \widetilde{H}_1) = (1, 2)_{-\frac{1}{2}}$	$H_1' = (H_1', \widetilde{H}_1') = (1, \bar{2})_{\frac{1}{2}}$
multiplets	$H_2 = (H_2, \widetilde{H}_2) = (1, \bar{2})_{\frac{1}{2}}$	$H_2' = (H_2', \widetilde{H}_2') = (1, 2)_{-\frac{1}{2}}$
Vector	$g = (g, \widetilde{g}) = (8, 1)_0$	$\Phi_g = (\phi_g, \psi_g) = (8, 1)_0$
multiplets	$W = (W, \widetilde{W}) = (1, 3)_0$	$\Phi_W = (\phi_W, \psi_W) = (1, 3)_0$
	$B = (B, \widetilde{B}) = (1, 1)_0$	$\Phi_B = (\phi_B, \psi_B) = (1, 1)_0$

For the above particle content, and imposing the full $U(2)_R$ on the superpotential, the theory is scale invariant and is given by the superpotential (after phase redefinitions)

$$W/\sqrt{2} = g'(\frac{1}{6}Q'\Phi_B Q - \frac{2}{3}U'\Phi_B U + \frac{1}{3}D'\Phi_B D - \frac{1}{2}L'\Phi_B L + E'\Phi_B E$$

$$-\frac{1}{2}H_1'\Phi_B H_1 + \frac{1}{2}H_2'\Phi_B H_2)$$
$$+ g_2(Q'\Phi_W Q + L'\Phi_W L + H_1'\Phi_W H_1 + H_2'\Phi_W H_2)$$
$$+ g_3(Q'\Phi_g Q + U'\Phi_g U + D'\Phi_g D). \tag{14.8}$$

After substitution in the Lagrangian (14.4), the superpotential (14.8) gives rise in the usual manner to gauge-quartic and gauge-Yukawa interactions. All interactions are gauge interactions! Table 14.1 and the superpotential (14.8) define the MN2SSM (in the supersymmetric limit). As expected, one missing ingredient in the vector-like superpotential (14.8) is chiral Yukawa terms. This is the fermion mass problem of extended supersymmetry frameworks.

The fermion mass problem in these models has many facets. First and foremost, the generation of any chiral spectrum must be a result of supersymmetry breaking. Secondly, the two sectors have to be distinguished with sufficiently heavy mirror fermions and (relatively) light ordinary fermions, with any mixing between the two sectors suppressed, at least in the case of the first two families. In addition there are the issues of the heavy third family and of the very light neutrinos in the ordinary sector, and subsequently, of flavor symmetries and their relation to supersymmetry breaking. One possible avenue to address these issues is to utilize the HSB Yukawa couplings of the previous section (in the framework of low-energy supersymmetry breaking) [170]. We note that unlike in the ($N = 1$) case with high-energy supersymmetry breaking, it is possible in this case that only one Higgs doublet acquires a *vev*. this is because HSB Yukawa terms are not necessarily constrained by holomorphicity.

These and other technical challenges in embedding the SM in a $N = 2$ framework are intriguing, but will not be pursued here. The $N = 2$ framework outlined here was originally proposed in Ref. [176], and more recently in Ref. [170]. Experimental status of models with three additional families was addressed in Ref. [15].

References

1. S. Weinberg: *The Quantum Theory of Fields. Vol. 3: Supersymmetry* (Cambridge University Press, Cambridge 2000)
2. N. Polonsky: "Essential Supersymmetry". In: The proceedings of TASI-98, *Neutrinos in Physics and Astrophysics*, ed. P. Langacker (World Scientific, Singapore 2000) pp. 94 - 200
3. G. Altarelli: "The standard electroweak theory and beyond," hep-ph/9811456. In: The proceedings of TASI-98, *Neutrinos in Physics and Astrophysics*, ed. P. Langacker (World Scientific, Singapore 2000) pp. 27 - 93
4. Y. Fukuda *et al.* [Super-Kamiokande Collaboration]: "Evidence for oscillation of atmospheric neutrinos," Phys. Rev. Lett. **81**, 1562 (1998)
5. P. Langacker: "Overview of Neutrino Physics and Astrophysics". In: The proceedings of TASI-98, *Neutrinos in Physics and Astrophysics*, ed. P. Langacker (World Scientific, Singapore 2000) pp. 1 - 26
6. V. Barger: "Neutrino masses and mixing at the millennium," hep-ph/0003212; "Overview of neutrino oscillation physics," hep-ph/0005011
7. The 1998 Review of Particle Physics, C. Caso *et al.* [Particle Data Group]: The Eur. Phys. J. C **3**, 1 (1998); The 2000 Review of Particle Physics, D.E. Groom *et al.* [Particle Data Group]: The Eur. Phys. J. C **15**, 1 (2000); http://pdg.lbl.gov/pdg.html
8. R.N. Mohapatra and J.C. Pati: "Left-Right gauge symmetry and an 'isoconjugate' model of CP violation," Phys. Rev. D **11**, 566 (1975)
9. R.S. Chivukula: "Models of electroweak symmetry breaking: Course," hep-ph/9803219
10. K. Lane: "Technicolor 2000," hep-ph/0007304
11. C. Quigg: "Electroweak symmetry breaking and the Higgs sector," hep-ph/9905369
12. R.S. Chivukula, B.A. Dobrescu, H. Georgi and C.T. Hill: "Top quark seesaw theory of electroweak symmetry breaking," Phys. Rev. D **59**, 075003 (1999)
13. For example, see J. Erler and P. Langacker: "Electroweak model and constraints on new physics". In: The Review of Particle Physics [7]
14. M.E. Peskin and J.D. Wells: "How can a heavy Higgs boson be consistent with the precision electroweak measurements?," hep-ph/0101342
15. H. He, N. Polonsky and S. Su: "Extra families, Higgs spectrum and oblique corrections," hep-ph/0102144, Phys. Rev. D, in press
16. T. Appelquist, J. Terning and L.C. Wijewardhana: "Postmodern technicolor," Phys. Rev. Lett. **79**, 2767 (1997)
17. N. Arkani-Hamed, S. Dimopoulos and G. Dvali: "Phenomenology, astrophysics and cosmology of theories with submillimeter dimensions and TeV scale quantum gravity," Phys. Rev. D **59**, 086004 (1999)

18. H.P. Nilles: "On the Low-energy limit of string and M theory," hep-ph/0004064. In: The proceedings of TASI-97, *Supersymmetry, Supergravity and Supercolliders*, ed. J.A. Bagger (World Scientific, Singapore 1999) p. 709

19. K.R. Dienes: "A practical introduction to string theory, string model-building, and string phenomenology I: Ten dimensions." In: The proceedings of TASI-98, *Neutrinos in Physics and Astrophysics*, ed. P. Langacker (World Scientific, Singapore 2000) pp. 201 - 302

20. L. Randall and R. Sundrum: "A large mass hierarchy from a small extra dimension," Phys. Rev. Lett. **83**, 3370 (1999)

21. L. Randall and R. Sundrum: "An alternative to compactification," Phys. Rev. Lett. **83**, 4690 (1999)
    S.B. Giddings, E. Katz and L. Randall: "Linearized gravity in brane backgrounds," JHEP **0003**, 023 (2000)

22. K.R. Dienes, E. Dudas and T. Gherghetta: "Extra space-time dimensions and unification," Phys. Lett. B **436**, 55 (1998)

23. N. Arkani-Hamed and M. Schmaltz: "Hierarchies without symmetries from extra dimensions," Phys. Rev. D **61**, 033005 (2000)

24. J.L. Hewett: "Indirect collider signals for extra dimensions," Phys. Rev. Lett. **82**, 4765 (1999)
    H. Davoudiasl, J.L. Hewett and T.G. Rizzo: "Phenomenology of the Randall-Sundrum gauge hierarchy model," Phys. Rev. Lett. **84**, 2080 (2000); "Experimental probes of localized gravity: On and off the wall," Phys. Rev. D **63**, 075004 (2001)
    I. Antoniadis and K. Benakli: "Large dimensions and string physics in future colliders," Int. J. Mod. Phys. A **15**, 4237 (2000)

25. F.C. Adams, G.L. Kane, M. Mbonye and M.J. Perry: "Proton decay, black holes, and large extra dimensions," hep-ph/0009154

26. P.H. Frampton and C. Vafa: "Conformal approach to particle phenomenology," hep-th/9903226

27. C. Csaki, W. Skiba and J. Terning: "β-functions of orbifold theories and the hierarchy problem," Phys. Rev. D **61**, 025019 (2000)

28. For example, see:
    A. Pomarol and M. Quiros: "The Standard model from extra dimensions," Phys. Lett. B **438**, 255 (1998)
    I. Antoniadis, S. Dimopoulos, A. Pomarol and M. Quiros: "Soft masses in theories with supersymmetry breaking by TeV compactification," Nucl. Phys. B **544**, 503 (1999)

29. A.E. Nelson and M.J. Strassler: "A Realistic supersymmetric model with composite quarks," Phys. Rev. D **56**, 4226 (1997)
    N. Arkani-Hamed, M.A. Luty and J. Terning: "Composite quarks and leptons from dynamical supersymmetry breaking without messengers," Phys. Rev. D **58**, 015004 (1998)

30. Other proposals to strong electroweak dynamics within supersymmetry include:
    G.F Giudice and A. Kusenko: "A Strongly interacting phase of the minimal supersymmetric model," Phys. Lett. B **439**, 55 (1998)
    K. Choi and H.D. Kim: "Dynamical solution to the μ-problem at TeV scale," Phys. Rev. D **61**, 015010 (2000)
    M.A. Luty, J. Terning and A.K. Grant: "Electroweak symmetry breaking by strong supersymmetric dynamics at the TeV scale," Phys. Rev. D **63**, 075001 (2001)

31. N. Arkani-Hamed, H. Cheng, B.A. Dobrescu and L.J. Hall: "Self-breaking of the standard model gauge symmetry," Phys. Rev. D **62**, 096006 (2000)

32. F. Abe *et al.* [CDF Collaboration]: "Study of t anti-t production in p anti-p collisions using total transverse energy," Phys. Rev. Lett. **75**, 3997 (1995)
    S. Abachi *et al.* [D0 Collaboration]: "Observation of the top quark," Phys. Rev. Lett. **74**, 2632 (1995)
33. P. Langacker and M. Luo: "Implications of precision electroweak experiments for m_t, $\rho(0)$, $\sin^2 \theta_W$ and grand unification," Phys. Rev. D **44**, 817 (1991)
34. D. Zeppenfeld: "Collider physics," hep-ph/9902307. In: The proceedings of TASI-98, *Neutrinos in Physics and Astrophysics*, ed. P. Langacker (World Scientific, Singapore 2000) pp. 303 - 350
35. F. Halzen: "Lectures on neutrino astronomy: Theory and experiment," astro-ph/9810368. In: The proceedings of TASI-98, *Neutrinos in Physics and Astrophysics*, ed. P. Langacker (World Scientific, Singapore 2000) pp. 524 - 569
    V. Barger, F. Halzen, D. Hooper and C. Kao: "Indirect search for neutralino dark matter with high-energy neutrinos," hep-ph/0105182
36. M. Dine: "Supersymmetry phenomenology (with a broad brush)." *Lectures given at Theoretical Advanced Study Institute in Elementary Particle Physics (TASI-96): Fields, Strings, and Duality, Boulder, CO, 2-28 Jun 1996*, hep-ph/9612389
37. X. Tata: "What is supersymmetry and how do we find it?" *Lectures given at 9th Jorge Andre Swieca Summer School: Particles and Fields, Sao Paulo, Brazil, 16-28 Feb 1997*, hep-ph/9706307
38. M. Drees: "An Introduction to supersymmetry." *Lectures given at Inauguration Conference of the Asia Pacific Center for Theoretical Physics (APCTP), Seoul, Korea, 4-19 Jun 1996*, hep-ph/9611409
39. J. Wess and B. Zumino: "A Lagrangian model invariant under supergauge transformations," Phys. Lett. **49B**, 52 (1974); "Supergauge transformations in four-dimensions," Nucl. Phys. B **70** (1974) 39
40. H.P. Nilles: "Supersymmetry, supergravity and particle physics," Phys. Rept. **110** (1984) 1
41. H.P. Nilles: "Beyond the Standard Model." In: The proceedings of TASI-90, *Testing the Standard Model*, eds. M. Cvetic and P. Langacker (World Scientific, Singapore 1991) pp. 633-718
42. H.P. Nilles: "Minimal supersymmetric Standard Model and grand unification," In: The proceedings of TASI-93, *The building blocks of creation*, eds. S. Raby and T. Walker (World Scientific, Singapore 1994) pp. 291-346
43. D.R.T. Jones: "Introduction to supersymmetry." In: The proceedings of TASI-93, *The building blocks of creation*, eds. S. Raby and T. Walker (World Scientific, Singapore 1994) pp. 259-290
44. J.D. Lykken: "Introduction to supersymmetry." *Lectures given at Theoretical Advanced Study Institute in Elementary Particle Physics (TASI-96): Fields, Strings, and Duality, Boulder, CO, 2-28 Jun 1996*, hep-th/9612114
45. S.P. Martin: "A Supersymmetry primer," hep-ph/9709356
46. J. Wess and J. Bagger: *Supersymmetry and supergravity*, 2nd edn. (Princeton, NJ, 1991)
47. K. Intriligator and N. Seiberg: "Lectures on supersymmetric gauge theories and electric-magnetic duality," Nucl. Phys. Proc. Suppl. **45BC**, 1 (1996) hep-th/9509066
48. H.E. Haber and G.L. Kane: "The search for supersymmetry: Probing physics beyond the standard model," Phys. Rept. **117**, 75 (1985)
49. H.E. Haber, "Introductory low-energy supersymmetry," *Lectures given at Theoretical Advanced Study Institute in Elementary Particle Physics (TASI-92): From Black Holes and Strings to Particles, Boulder, CO, 3-28 Jun 1992*, hep-ph/9306207.

50. D.I. Kazakov: "Beyond the standard model (in search of supersymmetry)," hep-ph/0012288
51. S. Abel et al. [SUGRA Working Group Collaboration]: "Report of the SUGRA working group for run II of the Tevatron," hep-ph/0003154
    R. Culbertson et al.: "Low-scale and gauge-mediated supersymmetry breaking at the Fermilab Tevatron Run II," hep-ph/0008070
52. A. Djouadi et al. [MSSM Working Group Collaboration]: "The minimal supersymmetric standard model: Group summary report," hep-ph/9901246
53. P. Fayet: "About superpartners and the origins of the supersymmetric Standard Model," hep-ph/0104302
54. J.L. Feng, N. Polonsky and S. Thomas: "The light higgsino - gaugino window," Phys. Lett. B **370**, 95 (1996)
55. G.R. Farrar: "Status of light gaugino scenarios," Nucl. Phys. Proc. Suppl. **62**, 485 (1998), hep-ph/9710277; "Experimental and cosmological implications of light gauginos," hep-ph/9710395.
56. S. Raby and K. Tobe: "The Phenomenology of SUSY models with a gluino LSP," Nucl. Phys. B **539**, 3 (1999)
    A. Mafi and S. Raby: "An analysis of a heavy gluino LSP at CDF: The heavy gluino window," Phys. Rev. D **62**, 035003 (2000)
57. J. Polchinski and L. Susskind: "Breaking of supersymmetry at intermediate-energy," Phys. Rev. D **26**, 3661 (1982)
    H.P. Nilles, M. Serdnicki, and D. Wyler: "Constraints on the stability of mass hierarchies in supergravity," Phys. Lett. **124B**, 337 (1983)
    A. Lahanas: "Light singlet, gauge hierarchy and supergravity," Phys. Lett. **124B**, 341 (1983)
    U. Ellwagner: "Nonrenormalizable interactions from supergravity, quantum corrections and effective low-energy theories," Phys. Lett. **133B**, 187 (1983)
58. K. Inoue, A. Kakuto, H. Komatsu and S. Takeshita: "Low-energy parameters and particle masses in a supersymmetric grand unified model," Prog. Theor. Phys. **67**, 1889 (1982); "Aspects of grand unified models with softly broken supersymmetry," Prog. Theor. Phys. **68**, 927 (1982); Erratum ibid. **70**, 330 (1983); "Renormalization of supersymmetry breaking parameters revisited," Prog. Theor. Phys. **71**, 413 (1984)
59. L. Girardello and M.T. Grisaru: "Soft breaking of supersymmetry," Nucl. Phys. B **194**, 65 (1982)
60. L.J. Hall and L. Randall: "Weak scale effective supersymmetry," Phys. Rev. Lett. **65**, 2939 (1990)
61. F. Borzumati, G.R. Farrar, N. Polonsky and S. Thomas: "Soft Yukawa couplings," Nucl. Phys. B **555**, 53 (1999); "Fermion masses without Yukawa couplings," hep-ph/9805314; "Precision measurements at the Higgs resonance: A probe of radiative fermion masses," hep-ph/9712428
62. I. Jack and D.R.T. Jones: "Non-standard soft supersymmetry breaking," Phys. Lett. B **457**, 101 (1999); "Quasi-infra-red fixed points and renormalization group invariant trajectories for non-holomorphic soft supersymmetry breaking," Phys. Rev. D **61**, 095002 (2000)
63. V. Barger, G.F. Giudice and T. Han; "Some New Aspects Of Supersymmetry R Parity Violating Interactions," Phys. Rev. D **40**, 2987 (1989)
    H. Dreiner: "An introduction to explicit R-parity violation," hep-ph/9707435; Pramana **51**, 123 (1998)
    G. Bhattacharyya: "R-parity-violating supersymmetric Yukawa couplings: A mini-review," Nucl. Phys. Proc. Suppl. **52A**, 83 (1997)
    R. Barbier et al.: "Report of the group on the R-parity violation," hep-ph/9810232

B. Allanach *et al.*: "Searching for R-parity violation at Run-II of the Tevatron," hep-ph/9906224

64. G.R. Farrar and P. Fayet: "Phenomenology of the production, decay, and detection of new hadronic states associated with supersymmetry," Phys. Lett. **76B**, 575 (1978)

65. A.H. Chamseddine and H. Dreiner: "Anomaly free gauged R symmetry in local supersymmetry," Nucl. Phys. B **458**, 65 (1996)
D.J. Castano, D.Z. Freedman and C. Manuel: "Consequences of supergravity with gauged $U(1)_R$ symmetry," Nucl. Phys. B **461**, 50 (1996)
N. Kitazawa, N. Maru and N. Okada: "Dynamical supersymmetry breaking with gauged $U(1)_R$ symmetry," Phys. Rev. D **62**, 077701 (2000); "Models of dynamical supersymmetry breaking with gauged $U(1)_R$ symmetry," Nucl. Phys. B **586**, 261 (2000)

66. L.E. Ibanez and G.G. Ross: "Discrete gauge symmetries and the origin of baryon and lepton number conservation in supersymmetric versions of the standard model," Nucl. Phys. B **368**, 3 (1992)

67. D. Kapetanakis, P. Mayr and H.P. Nilles: "Discrete symmetries and solar neutrino mixing," Phys. Lett. B **282**, 95 (1992)

68. L.J. Hall and M. Suzuki: "Explicit R-parity breaking in supersymmetric models," Nucl. Phys. B **231**, 419 (1984)

69. M. Drees and M.M. Nojiri: "The Neutralino relic density in minimal N=1 supergravity," Phys. Rev. D **47**, 376 (1993)

70. J.D. Wells: "Supersymmetric dark matter with a cosmological constant," Phys. Lett. B **443**, 196 (1998)

71. R. Arnowitt and P. Nath, "Cosmological constraints and SU(5) supergravity grand unification," Phys. Lett. B **299**, 58 (1993); *Erratum-ibid.* B **307**, 58 (1993)
J.L. Feng, K.T. Matchev and F. Wilczek: "Prospects for indirect detection of neutralino dark matter," astro-ph/0008115

72. L. Covi, J.E. Kim and L. Roszkowski: "Axinos as cold dark matter," Phys. Rev. Lett. **82**, 4180 (1999)
E.J. Chun and H.B. Kim: "Nonthermal axino as cool dark matter in supersymmetric standard model without R-parity," Phys. Rev. D **60**, 095006 (1999)
L. Covi, H. Kim, J.E. Kim and L. Roszkowski: "Axinos as dark matter," hep-ph/0101009

73. J. Ellis, T. Falk, G. Ganis and K.A. Olive: "Supersymmetric dark matter in the light of LEP and the Tevatron collider," Phys. Rev. D **62**, 075010 (2000)
M. Drees, Y.G. Kim, M.M. Nojiri, D. Toya, K. Hasuko and T. Kobayashi: "Scrutinizing LSP dark matter at the LHC," Phys. Rev. D **63**, 035008 (2001)

74. F.M. Borzumati, M. Drees and M.M. Nojiri: Implications for supersymmetric dark matter detection from radiative b decays," Phys. Rev. D **51**, 341 (1995)
M. Drees, Y.G. Kim, T. Kobayashi and M.M. Nojiri: "Direct detection of neutralino dark matter and the anomalous dipole moment of the muon," hep-ph/0011359

75. X. Tata: "Supersymmetry: Where it is and how to find it," *Lectures given at Theoretical Advanced Study Institute in Elementary Particle Physics (TASI-95): QCD and Beyond, Boulder, CO, 4-30 Jun 1995*, hep-ph/9510287

76. A. Bottino and N. Fornengo: "Dark matter and its particle candidates," *Lectures given at 5th ICTP School on Nonaccelerator Astroparticle Physics, Trieste, Italy, 29 Jun - 10 Jul 1998*, hep-ph/9904469

77. M. Drees: "Particle dark matter physics: An update," Pramana **51**, 87 (1998), hep-ph/9804231

78. N. Polonsky and S. Su: "More corrections to the Higgs mass in supersymmetry," hep-ph/0010113, Phys. Lett. B, in press
79. K.R. Dienes, C. Kolda and J. March-Russell: "Kinetic mixing and the supersymmetric gauge hierarchy," Nucl. Phys. B **492**, 104 (1997)
   K.S. Babu, C. Kolda and J. March-Russell: "Implications of generalized $Z - Z'$ mixing," Phys. Rev. D **57**, 6788 (1998)
80. H. Georgi and S.L. Glashow: "Unity of all elementary particle forces," Phys. Rev. Lett. **32**, 438 (1974)
81. H. Georgi, H.R. Quinn and S. Weinberg: "Hierarchy of interactions in unified gauge theories," Phys. Rev. Lett. **33**, 451 (1974)
82. K.R. Dienes: "String theory and the path to unification: A Review of recent developments," Phys. Rept. **287**, 447 (1997)
83. L.E. Ibanez; "Recent developments in physics far beyond the standard model," hep-ph/9901292
   L.E. Ibanez, C. Munoz and S. Rigolin: "Aspects of type I string phenomenology," Nucl. Phys. B **553**, 43 (1999)
84. P. Langacker: "Physics implications of precision electroweak experiments," talk given at *LEP Fest 2000*, October 2000, CERN, hep-ph/0102085
85. P. Langacker and N. Polonsky: "Uncertainties in coupling constant unification," Phys. Rev. D **47**, 4028 (1993); 'The Bottom mass prediction in supersymmetric grand unification: Uncertainties and constraints," Phys. Rev. D **49**, 1454 (1994); "The Strong coupling, unification, and recent data," Phys. Rev. D **52**, 3081 (1995); N. Polonsky [90]
86. R. Hempfling: "Coupling constant unification in extended SUSY models," Phys. Lett. B **351**, 206 (1995)
   C. Kolda and J. March-Russell: "Low-energy signatures of semi-perturbative unification," Phys. Rev. D **55**, 4252 (1997)
   D. Ghilencea, M. Lanzagorta and G.G. Ross: "Unification predictions," Nucl. Phys. B **511**, 3 (1998); "Strong unification," Phys. Lett. B **415**, 253 (1997)
87. R. Arnowitt and P. Nath, "SUSY mass spectrum in SU(5) supergravity grand unification," Phys. Rev. Lett. **69**, 725 (1992); "Radiative breaking, proton stability and the viability of no scale supergravity models," Phys. Lett. B **287**, 89 (1992); "Predictions in SU(5) supergravity grand unification with proton stability and relic density constraints," Phys. Rev. Lett. **70**, 3696 (1993); "Supersymmetry and supergravity: Phenomenology and grand unification," hep-ph/9309277; "Supergravity unified models," hep-ph/9708254
   J. Hisano, H. Murayama and T. Yanagida: "Nucleon decay in the minimal supersymmetric SU(5) grand unification," Nucl. Phys. B **402**, 46 (1993)
   K. Hagiwara and Y. Yamada: "GUT threshold effects in supersymmetric SU(5) models," Phys. Rev. Lett. **70**, 709 (1993)
   V. Barger, M.S. Berger and P. Ohmann: "Supersymmetric grand unified theories: Two loop evolution of gauge and Yukawa couplings," Phys. Rev. D **47**, 1093 (1993)
   M. Carena, S. Pokorski and C.E. Wagner: "On the unification of couplings in the minimal supersymmetric Standard Model," Nucl. Phys. B **406**, 59 (1993)
   W.A. Bardeen, M. Carena, S. Pokorski and C.E. Wagner: "Infrared fixed point solution for the top quark mass and unification of couplings in the MSSM," Phys. Lett. B **320**, 110 (1994)
   L.J. Hall, R. Rattazzi and U. Sarid: "The Top quark mass in supersymmetric SO(10) unification," Phys. Rev. D **50**, 7048 (1994)
   P.H. Chankowski, Z. Pluciennik and S. Pokorski: "$\sin^2 \theta_W (M(Z))$ in the MSSM and unification of couplings," Nucl. Phys. B **439**, 23 (1995)
   D.M. Pierce, J.A. Bagger, K. Matchev and R. Zhang, "Precision corrections

in the minimal supersymmetric standard model," Nucl. Phys. **B491**, 3 (1997); "Gauge and Yukawa unification in models with gauge mediated supersymmetry breaking," Phys. Rev. Lett. **78**, 1002 (1997)

Z. Berezhiani, Z. Tavartkiladze and M. Vysotsky: "d = 5 operators in SUSY GUT: Fermion masses versus proton decay," hep-ph/9809301

T. Goto and T. Nihei: "Effect of RRRR dimension five operator on the proton decay in the minimal SU(5) SUGRA GUT model," Phys. Rev. D **59**, 115009 (1999)

R. Dermisek, A. Mafi and S. Raby: "SUSY GUTs under siege : Proton decay," Phys. Rev. D **63**, 035001 (2001)

Y. Kawamura: "Triplet-doublet splitting, proton stability and extra dimension," hep-ph/0012125

88. R.N. Mohapatra: "Supersymmetric grand unification," *Lectures given at the Theoretical Advanced Study Institute in Elementary Particle Physics (TASI-97): Supersymmetry, Supergravity and Supercolliders, Boulder, CO, 1-7 Jun 1997*, hep-ph/9801235; "Supersymmetric grand unification: An update," hep-ph/9911272.

89. M.S. Chanowitz, J. Ellis and M.K. Gaillard: "The price of natural flavor conservation in neutral weak interactions," Nucl. Phys. B **128**, 506 (1977)

A.J. Buras, J. Ellis, M.K. Gaillard and D.V. Nanopoulos: "Aspects of the grand unification of strong, weak and electromagnetic interactions," Nucl. Phys. B **135**, 66 (1978)

90. N. Polonsky: "On supersymmetric b - tau unification, gauge unification, and fixed points," Phys. Rev. D **54**, 4537 (1996)

91. Y. Totsuka [Super-Kamiokande Collaboration]: Unpublished talk presented in *Supersymmetry 2000*, CERN, June 2000.

92. N. Polonsky: "Origins and renormalization of the superparticle spectrum," Nucl. Phys. Proc. Suppl. **62**, 204 (1998)

93. C.T. Hill: "Quark and lepton masses from renormalization group fixed points," Phys. Rev. D **24**, 691 (1981)

94. B. Pendleton and G.G. Ross: "Mass and mixing angle predictions from infrared fixed points," Phys. Lett. **98B**, 291 (1981)

95. L.E. Ibanez and G.G. Ross: "low-energy predictions in supersymmetric grand unified theories," Phys. Lett. **105B**, 439 (1981)

L.E. Ibanez and C. Lopez: "N=1 Supergravity, the breaking of $SU(2) \times U(1)$ and the top quark mass," Phys. Lett. **126B**, 54 (1983)

L.E. Ibanez, C. Lopez and C. Munoz: "The low-energy supersymmetric spectrum according to $N = 1$ supergravity GUTs," Nucl. Phys. B **256**, 218 (1985)

H.P. Nilles, "Dynamically broken supergravity and the hierarchy problem," Phys. Lett. **115B**, 193 (1982)

K. Inoue *et al.* [58]

P.H. Chankowski: "Radiative $SU(2) \times U(1)$ breaking in the supersymmetric standard model and decoupling of heavy squarks and gluino," Phys. Rev. D **41**, 2877 (1990)

M. Drees and M.M. Nojiri, "Radiative symmetry breaking in minimal N=1 supergravity with large Yukawa couplings," Nucl. Phys. B **369**, 54 (1992)

96. J.M. Frere, D.R.T. Jones and S. Raby: "Fermion masses and induction of the weak scale by supergravity," Nucl. Phys. B **222**, 11 (1983)

L. Alvarez-Gaume, J. Polchinski and M.B. Wise: "Minimal low-energy supergravity," Nucl. Phys. B **221**, 495 (1983)

C. Kounnas, A.B. Lahanas, D.V. Nanopoulos and M. Quiros: "Low-energy behavior of realistic locally supersymmetric grand unified theories," Nucl. Phys. B **236**, 438 (1984)

J.P. Derendinger and C.A. Savoy: "Quantum effects and $SU(2) \times U(1)$ breaking in supergravity gauge theories," Nucl. Phys. B **B37**, 307 (1984)

97. J.L. Feng, C. Kolda and N. Polonsky: "Solving the supersymmetric flavor problem with radiatively generated mass hierarchies," Nucl. Phys. B **546**, 3 (1999)

J. Bagger, J.L. Feng and N. Polonsky: "Naturally heavy scalars in supersymmetric grand unified theories," Nucl. Phys. **B563**, 3 (1999)

J.A. Bagger, J.L. Feng, N. Polonsky and R. Zhang: "Superheavy supersymmetry from scalar mass A-parameter fixed points," Phys. Lett. B **473**, 264 (2000)

98. J.L. Feng and T. Moroi: "Supernatural supersymmetry: Phenomenological implications of anomaly-mediated supersymmetry breaking," Phys. Rev. D **61**, 095004 (2000)

J.L. Feng, K.T. Matchev and T. Moroi: "Multi-TeV scalars are natural in minimal supergravity," Phys. Rev. Lett. **84**, 2322 (2000); "Focus points and naturalness in supersymmetry," Phys. Rev. D **61**, 075005 (2000)

99. S.P. Martin and M.T. Vaughn: "Two loop renormalization group equations for soft supersymmetry breaking couplings," Phys. Rev. D **50**, 2282 (1994)

Y. Yamada: "Two loop renormalization group equations for soft SUSY breaking scalar interactions: Supergraph method," Phys. Rev. D **50**, 3537 (1994)

I. Jack and D.R.T. Jones: "Soft supersymmetry breaking and finiteness," Phys. Lett. B **333**, 372 (1994)

I. Jack, D.R.T. Jones, S.P. Martin, M.T. Vaughn and Y. Yamada: "Decoupling of the epsilon scalar mass in softly broken supersymmetry," Phys. Rev. D **50**, 5481 (1994)

100. R. Arnowitt and P. Nath, "Loop corrections to radiative breaking of electroweak symmetry in supersymmetry," Phys. Rev. D **46**, 3981 (1992)

S. Kelley, J.L. Lopez, D.V. Nanopoulos, H. Pois and K. Yuan, "Aspects of radiative electroweak breaking in supergravity models," Nucl. Phys. B **398**, 3 (1993)

V. Barger, M.S. Berger and P. Ohmann: "The supersymmetric particle spectrum," Phys. Rev. D **49**, 4908 (1994)

G.L. Kane, C. Kolda, L. Roszkowski and J.D. Wells: "Study of constrained minimal supersymmetry," Phys. Rev. D **49**, 6173 (1994)

M. Olechowski and S. Pokorski, "Bottom - up approach to unified supergravity models," Nucl. Phys. B **404**, 590 (1993)

M. Carena, M. Olechowski, S. Pokorski and C.E. Wagner: "Radiative electroweak symmetry breaking and the infrared fixed point of the top quark mass," Nucl. Phys. B **419**, 213 (1994); "Electroweak symmetry breaking and bottom - top Yukawa unification," Nucl. Phys. B **426**, 269 (1994)

N. Polonsky and A. Pomarol: "GUT effects in the soft supersymmetry breaking terms," Phys. Rev. Lett. **73**, 2292 (1994); "Nonuniversal GUT corrections to the soft terms and their implications in supergravity models," Phys. Rev. D **51**, 6532 (1995)

S.A. Abel and B. Allanach, "The quasifixed MSSM," Phys. Lett. B **415**, 371 (1997)

M. Carena, P. Chankowski, M. Olechowski, S. Pokorski and C.E. Wagner: "Bottom - up approach and supersymmetry breaking," Nucl. Phys. B **491**, 103 (1997)

101. J.E. Kim and H.P. Nilles: "The μ-problem and the strong CP problem," Phys. Lett. **138B**, 150 (1984); "Symmetry principles toward solutions of the μ-problem," Mod. Phys. Lett. A **9**, 3575 (1994)

102. G.F. Giudice and A. Masiero: "A natural solution to the μ-problem in supergravity theories," Phys. Lett. B **206**, 480 (1988)
103. C. Kolda, S. Pokorski and N. Polonsky: "Stabilized singlets in supergravity as a source of the μ-parameter," Phys. Rev. Lett. **80**, 5263 (1998)
104. N. Polonsky: "The μ-parameter of supersymmetry," hep-ph/9911329
105. R. Barbieri and L. Maiani: "Renormalization of the electroweak ρ-parameter from supersymmetric particles," Nucl. Phys. B **224**, 32 (1983)
106. H.E. Haber and A. Pomarol: "Constraints from global symmetries on radiative corrections to the Higgs sector," Phys. Lett. B **302**, 435 (1993)
107. L. Durand and J. L. Lopez: "Upper bounds on Higgs and top quark masses in the flipped $SU(5) \times U(1)$ superstring model," Phys. Lett. B **217**, 463 (1989)
     M. Drees: "Supersymmetric models with extended Higgs sector," Int. J. Mod. Phys. A **4**, 3635 (1989)
108. S. Abel, S. Sarkar and P. White: "On the cosmological domain wall problem for the minimally extended supersymmetric standard model," Nucl. Phys. B **454**, 663 (1995)
     U. Ellwanger, M. Rausch de Traubenberg and C.A. Savoy: "Phenomenology of supersymmetric models with a singlet," Nucl. Phys. B **492**, 21 (1997)
     S.J. Huber and M.G. Schmidt: "Electroweak baryogenesis: Concrete in a SUSY model with a gauge singlet," hep-ph/0003122
109. M. Cvetic and P. Langacker: "Z' physics and supersymmetry," hep-ph/9707451, and references therein
110. P. Langacker and J. Wang: "$U(1)'$ symmetry breaking in supersymmetric $E(6)$ models," Phys. Rev. D **58**, 115010 (1998)
111. J.R. Espinosa and M. Quiros; "Gauge unification and the supersymmetric light Higgs mass," Phys. Rev. Lett. **81**, 516 (1998)
112. P. Langacker, N. Polonsky and J. Wang: "A low-energy solution to the μ-problem in gauge mediation," Phys. Rev. D **60**, 115005 (1999)
113. P. Igo-Kemenes [LEP Higgs Working Group Collaboration]: Talk presented to LEPC, CERN, November 2000; A.N. Okpara [LEP Collaborations]: hep-ph/0105151
114. H.E. Haber: 'Low-energy supersymmetry and its phenomenology," hep-ph/0103095
115. H.E. Haber: "Higgs boson masses and couplings in the minimal supersymmetric model," hep-ph/9707213
116. M.A. Diaz and H.E. Haber: "Can the Higgs mass be entirely due to radiative corrections?," Phys. Rev. D **46**, 3086 (1992)
     H.E. Haber and R. Hempfling: "Can the mass of the lightest Higgs boson of the minimal supersymmetric model be larger than M_Z?," Phys. Rev. Lett. **66**, 1815 (1991); "The renormalization group improved Higgs sector of the minimal supersymmetric model," Phys. Rev. D **48**, 4280 (1993)
     H.E. Haber, R. Hempfling and A.H. Hoang: "Approximating the radiatively corrected Higgs mass in the minimal supersymmetric model," Z. Phys. C **75**, 539 (1997)
     Y. Okada, M. Yamaguchi and T. Yanagida: "Upper bound of the lightest Higgs boson mass in the minimal supersymmetric standard model," Prog. Theor. Phys. **85**, 1 (1991)
     J. Ellis, G. Ridolfi and F. Zwirner: "Radiative corrections to the masses of supersymmetric Higgs bosons," Phys. Lett. B **257**, 83 (1991); "On radiative corrections to supersymmetric Higgs boson masses and their implications for LEP searches," Phys. Lett. B **262**, 477 (1991)
     R. Barbieri, M. Frigeni and F. Caravaglios: "The supersymmetric Higgs for heavy superpartners," Phys. Lett. B **258**, 167 (1991)

J.R. Espinosa and M. Quiros: "Two loop radiative corrections to the mass of the lightest Higgs boson in supersymmetric standard models," Phys. Lett. B **266**, 389 (1991)

J.R. Espinosa and M. Quiros: "Higgs triplets in the supersymmetric standard model," Nucl. Phys. B **384**, 113 (1992)

V. Barger, M.S. Berger, P. Ohmann and R.J. Phillips: "Phenomenological implications of the m_t RGE fixed point for SUSY Higgs boson searches," Phys. Lett. B **314**, 351 (1993)

P. Langacker and N. Polonsky: "Implications of Yukawa unification for the Higgs sector in supersymmetric grand unified models," Phys. Rev. D **50**, 2199 (1994)

M. Carena, J.R. Espinosa, M. Quiros and C.E. Wagner: "Analytical expressions for radiatively corrected Higgs masses and couplings in the MSSM," Phys. Lett. B **355**, 209 (1995)

M. Carena, M. Quiros and C.E. Wagner: "Effective potential methods and the Higgs mass spectrum in the MSSM," Nucl. Phys. **B461**, 407 (1996)

M. Carena, P.H. Chankowski, S. Pokorski and C.E. Wagner: "The Higgs boson mass as a probe of the minimal supersymmetric standard model," Phys. Lett. B **441**, 205 (1998)

H.E. Haber: "How well can we predict the mass of the Higgs boson of the minimal supersymmetric model?," hep-ph/9901365

J.R. Espinosa and R. Zhang: "Complete two-loop dominant corrections to the mass of the lightest CP-even Higgs boson in the minimal supersymmetric standard model," Nucl. Phys. B **586**, 3 (2000)

M. Carena, H.E. Haber, S. Heinemeyer, W. Hollik, C.E. Wagner and G. Weiglein: "Reconciling the two-loop diagrammatic and effective field theory computations of the mass of the lightest CP-even Higgs boson in the MSSM," Nucl. Phys. B **580**, 29 (2000)

117. D.M. Pierce: "Renormalization of supersymmetric theories," *Lectures given at the Theoretical Advanced Study Institute in Elementary Particle Physics (TASI-97): Supersymmetry, Supergravity and Supercolliders, Boulder, CO, 1-7 Jun 1997*, hep-ph/9805497

A. Djouadi, P. Gambino, S. Heinemeyer, W. Hollik, C. Junger and G. Weiglein: "Supersymmetric contributions to electroweak precision observables: QCD corrections," Phys. Rev. Lett. **78**, 3626 (1997); "Leading QCD corrections to scalar quark contributions to electroweak precision observables," Phys. Rev. D **57**, 4179 (1998)

J. Erler and D.M. Pierce: "Bounds on supersymmetry from electroweak precision analysis," Nucl. Phys. B **526**, 53 (1998)

G.C. Cho and K. Hagiwara: "Supersymmetry versus precision experiments revisited," Nucl. Phys. B **574**, 623 (2000)

118. J.A. Bagger: "Weak scale supersymmetry: Theory and practice," *Lectures given at Theoretical Advanced Study Institute in Elementary Particle Physics (TASI-95): QCD and Beyond, Boulder, CO, 4-30 Jun 1995*, hep-ph/9604232

119. S. Dawson: "The MSSM and why it works," *Lectures given at the Theoretical Advanced Study Institute in Elementary Particle Physics (TASI-97): Supersymmetry, Supergravity and Supercolliders, Boulder, CO, 1-7 Jun 1997*, hep-ph/9712464

120. J. Ellis, S. Ferrara and D.V. Nanopoulos, "CP violation and supersymmetry," Phys. Lett. **114B**, 231 (1982)

W. Buchmuller and D. Wyler: "CP violation and R invariance in supersymmetric models of strong and electroweak interactions," Phys. Lett. **121B**, 321 (1983)

J. Polchinski and M.B. Wise: "The electric dipole moment of the neutron in low-energy supergravity," phys. lett. **125B**, 393 (1983)

W. Fischler, S. Paban and S. Thomas: "Bounds on microscopic physics from P and T violation in atoms and molecules," Phys. Lett. B **289**, 373 (1992)

Y. Kizukuri and N. Oshimo: "The Neutron and electron electric dipole moments in supersymmetric theories," Phys. Rev. D **46**, 3025 (1992)

S. Bertolini and F. Vissani: "On soft breaking and CP phases in the supersymmetric standard model," Phys. Lett. B **324**, 164 (1994)

S. Dimopoulos and S. Thomas: "Dynamical relaxation of the supersymmetric CP violating phases," Nucl. Phys. B **465**, 23 (1996)

T. Moroi: "Electric dipole moments in gauge mediated models and a solution to the SUSY CP problem," Phys. Lett. B **447**, 75 (1999)

121. E. Accomando, R. Arnowitt and B. Dutta: "Grand unification scale CP violating phases and the electric dipole moment," Phys. Rev. D **61**, 115003 (2000)

122. T. Ibrahim and P. Nath: "The Neutron and the lepton EDMs in MSSM, large CP violating phases, and the cancellation mechanism," Phys. Rev. D **58**, 111301 (1998); "Slepton flavor nonuniversality, the muon EDM and its proposed sensitive search at Brookhaven," hep-ph/0105025

M. Brhlik, G.J. Good and G.L. Kane: "Electric dipole moments do not require the CP violating phases of supersymmetry to be small," Phys. Rev. D **59**, 115004 (1999)

123. J.A. Bagger, K.T. Matchev and R. Zhang: "QCD corrections to flavor changing neutral currents in the supersymmetric standard model," Phys. Lett. B **412**, 77 (1997)

M. Ciuchini et al., "$\Delta M(K)$ and $\epsilon(K)$ in SUSY at the next-to-leading order," JHEP **10**, 008 (1998)

124. J.A. Grifols and A. Mendez: "Constraints on supersymmetric particle masses from $(g-2)_\mu$," Phys. Rev. D **26**, 1809 (1982)

J. Ellis, J. Hagelin and D.V. Nanopoulos: "Spin 0 leptons and the anomalous magnetic moment of the muon," Phys. Lett. **116B**, 283 (1982)

R. Barbieri and L. Maiani: "The muon anomalous magnetic moment in broken supersymmetric theories," Phys. Lett. **117B**, 203 (1982)

D.A. Kosower, L.M. Krauss and N. Sakai: "Low-energy supergravity and the anomalous magnetic moment of the muon," Phys. Lett. **133B**, 305 (1983)

125. A. Czarnecki and W.J. Marciano: "The muon anomalous magnetic moment: A harbinger for 'new physics'," hep-ph/0102122

126. S. Bertolini, F. Borzumati and A. Masiero: "New constraints on squark and gluino masses from radiative B decays," Phys. Lett. B **192**, 437 (1987)

S. Bertolini, F. Borzumati, A. Masiero and G. Ridolfi: "Effects of supergravity induced electroweak breaking on rare B decays and mixings," Nucl. Phys. B **353**, 591 (1991)

F.M. Borzumati: "The Decay $b \to s\gamma$ in the MSSM revisited," Z. Phys. C **63**, 291 (1994)

127. F. Gabbiani, E. Gabrielli, A. Masiero and L. Silvestrini: "A complete analysis of FCNC and CP constraints in general SUSY extensions of the standard model," Nucl. Phys. B **477**, 321 (1996)

A. Masiero and L. Silvestrini: "Two lectures on FCNC and CP violation in supersymmetry," hep-ph/9711401

R. Contino and I. Scimemi: "The supersymmetric flavor problem for heavy first two generation scalars at next-to-leading order," Eur. Phys. J. C **10**, 347 (1999)

A. Masiero and H. Murayama: "Can ϵ'/ϵ be supersymmetric?," hep-ph/9903363

A.J. Buras, P. Gambino, M. Gorbahn, S. Jager and L. Silvestrini: "ϵ'/ϵ and rare K and B decays in the MSSM," Nucl. Phys. B **592**, 55 (2001)

R. Barbieri, L. Hall and A. Strumia: "Violations of lepton flavor and CP in supersymmetric unified theories," Nucl. Phys. B **445**, 219 (1995)

P. Ciafaloni, A. Romanino and A. Strumia: "Lepton flavor violations in SO(10) with large tan β," Nucl. Phys. B **458**, 3 (1996)

F. Borzumati and A. Masiero, "Large muon and electron number violations in supergravity theories," Phys. Rev. Lett. **57**, 961 (1986)

J. Hisano, T. Moroi, K. Tobe and M. Yamaguchi: "Lepton flavor violation via right-handed neutrino Yukawa couplings in supersymmetric standard model," Phys. Rev. D **53**, 2442 (1996)

B. de Carlos, J.A. Casas and J.M. Moreno; "Constraints on supersymmetric theories from $\mu \to e\gamma$," Phys. Rev. D **53**, 6398 (1996)

U. Chattopadhyay and P. Nath: "Probing supergravity grand unification in the Brookhaven $g - 2$ experiment," Phys. Rev. D **53**, 1648 (1996);

T. Moroi: "The Muon anomalous magnetic dipole moment in the minimal supersymmetric standard model," Phys. Rev. D **53**, 6565 (1996); *Erratum ibid.* D **56**, 4424 (1997)

M. Carena, G.F. Giudice and C.E. Wagner: "Constraints on supersymmetric models from the muon anomalous magnetic moment," Phys. Lett. B **390**, 234 (1997)

J.L. Feng and K.T. Matchev: "Supersymmetry and the anomalous magnetic moment of the muon," hep-ph/0102146

S.P. Martin and J.D. Wells: "Muon anomalous magnetic dipole moment in supersymmetric theories," hep-ph/0103067

J. Hisano and K. Tobe: "Neutrino masses, muon $g - 2$, and lepton-flavour violation in the supersymmetric see-saw model," hep-ph/0102315

S. Baek, T. Goto, Y. Okada and K. Okumura: "Muon anomalous magnetic moment, lepton flavor violation, and flavor changing neutral current processes in SUSY GUT with right-handed neutrino," hep-ph/0104146

M. Graesser and S. Thomas: "Supersymmetric Relations Among Electromagnetic Dipole Operators," hep-ph/0104254

Z. Chacko and G.D. Kribs: "Constraints on lepton flavor violation in the MSSM from the muon anomalous magnetic moment measurement," hep-ph/0104317

128. K. Choi, J.S. Lee and C. Munoz: "Supergravity radiative effects on soft terms and the μ-term," Phys. Rev. Lett. **80**, 3686 (1998)

129. H.P. Nilles and N. Polonsky: "Gravitational divergences as a mediator of supersymmetry breaking," Phys. Lett. B **412**, 69 (1997)

N. Polonsky: "Supergravity miracles: Phenomenology with gravitational divergences," hep-ph/9809422

130. K. Choi, J.E. Kim and H.P. Nilles: "Cosmological constant and soft terms in supergravity," Phys. Rev. Lett. **73**, 1758 (1994)

131. V.S. Kaplunovsky and J. Louis: "Model independent analysis of soft terms in effective supergravity and in string theory," Phys. Lett. B **306**, 269 (1993)

A. Brignole, L.E. Ibanez, C. Munoz and C. Scheich: "Some issues in soft SUSY breaking terms from dilaton / moduli sectors," Z. Phys. C **74**, 157 (1997)

Y. Kawamura and T. Kobayashi: "Generic formula of soft scalar masses in string models," Phys. Rev. D **56**, 3844 (1997)

H.P. Nilles, M. Olechowski and M. Yamaguchi: "Supersymmetry breaking and soft terms in M-theory," Phys. Lett. B **415**, 24 (1997)

132. S. Dimopoulos and H. Georgi: "Softly broken supersymmetry and $SU(5)$," Nucl. Phys. B **193**, 150 (1981)

R. Barbieri, S. Ferrara and C.A. Savoy: "Gauge models with spontaneously broken local supersymmetry," Phys. Lett. **119B**, 343 (1982)

P. Nath, R. Arnowitt and A.H. Chamseddine: "Gravity induced symmetry breaking and ground state of local supersymmetric GUTs," Phys. Lett. **121B**, 33 (1983)

H.P. Nilles: "Supergravity generates hierarchies," Nucl. Phys. B **217**, 366 (1983)

H.P. Nilles, M. Srednicki and D. Wyler; "Weak interaction breakdown induced by supergravity," Phys. Lett. **120B**, 346 (1983)

L. Hall, J. Lykken and S. Weinberg; "Supergravity as the messenger of super-symmetry breaking," Phys. Rev. D **27**, 2359 (1983)

133. L. Randall and R. Sundrum: "Out of this world supersymmetry breaking," Nucl. Phys. B **557**, 79 (1999)

G.F. Giudice, M.A. Luty, H. Murayama and R. Rattazzi: "Gaugino mass without singlets," JHEP **9812**, 027 (1998)

I. Jack and D.R.T. Jones: "RG invariant solutions for the soft supersymmetry breaking parameters," Phys. Lett. B **465**, 148 (1999)

A. Pomarol and R. Rattazzi: "Sparticle masses from the superconformal anomaly," JHEP **9905**, 013 (1999)

J.A. Bagger, T. Moroi and E. Poppitz: "Anomaly mediation in supergravity theories," JHEP **0004**, 009 (2000)

134. G.D. Kribs: "Disrupting the one loop renormalization group invariant M/α in supersymmetry," Nucl. Phys. B **535**, 41 (1998)

135. K. Choi, N. Polonsky, and Y. Shirman: Unpublished

M.K. Gaillard and B. Nelson: "Quantum induced soft supersymmetry breaking in supergravity," Nucl. Phys. B **588**, 197 (2000)

136. D.E. Kaplan and G.D. Kribs: "Gaugino-assisted anomaly mediation," JHEP **0009**, 048 (2000)

137. M. Dine, A.E. Nelson and Y. Shirman: "Low-energy dynamical supersymmetry breaking simplified," Phys. Rev. D **51**, 1362 (1995)

M. Dine, A.E. Nelson, Y. Nir and Y. Shirman: "New tools for low-energy dynamical supersymmetry breaking," Phys. Rev. D **53**, 2658 (1996)

J.A. Bagger, K. Matchev, D.M. Pierce and R. Zhang: "Weak scale phenomenology in models with gauge mediated supersymmetry breaking," Phys. Rev. D **55**, 3188 (1997)

S. Dimopoulos, S. Thomas and J.D. Wells: "Sparticle spectroscopy and electroweak symmetry breaking with gauge mediated supersymmetry breaking," Nucl. Phys. B **488**, 39 (1997)

S.P. Martin: "Generalized messengers of supersymmetry breaking and the sparticle mass spectrum," Phys. Rev. D **55**, 3177 (1997)

S. Dimopoulos and G.F. Giudice: "Multi-messenger theories of gauge-mediated supersymmetry breaking," Phys. Lett. B **393**, 72 (1997)

G.F. Giudice and R. Rattazzi: "Theories with gauge mediated supersymmetry breaking," Phys. Rept. **322**, 419 (1999)

138. A.G. Cohen, D.B. Kaplan and A.E. Nelson: "Counting 4π's in strongly coupled supersymmetry," Phys. Lett. B **412**, 301 (1997)

139. S. Raby and K. Tobe: "Dynamical SUSY breaking with a hybrid messenger sector," Phys. Lett. B **437**, 337 (1998)

140. M. Dine and A.E. Nelson: "Dynamical supersymmetry breaking at low-energies," Phys. Rev. D **48**, 1277 (1993)

G. Dvali, G.F. Giudice and A. Pomarol: "The μ-problem in theories with gauge mediated supersymmetry breaking," Nucl. Phys. B **478**, 31 (1996)

T. Yanagida: "A Solution to the μ-problem in gauge mediated supersymmetry breaking models," Phys. Lett. B **400**, 109 (1997)

H.P. Nilles and N. Polonsky [129]

K. Agashe and M. Graesser: "Improving the fine tuning in models of low-energy gauge mediated supersymmetry breaking," Nucl. Phys. B **507**, 3 (1997)

A. de Gouvea, A. Friedland and H. Murayama: "Next-to-minimal supersymmetric standard model with the gauge mediation of supersymmetry breaking," Phys. Rev. D **57**, 5676 (1998)

E.J. Chun: "Strong CP and μ-problem in theories with gauge mediated supersymmetry breaking," Phys. Rev. D **59**, 015011 (1999)

P. Langacker, N. Polonsky and J. Wang [112]

T. Han, D. Marfatia and R. Zhang: "A gauge-mediated supersymmetry breaking model with an extra singlet Higgs field," Phys. Rev. D **61**, 013007 (2000)

N. Polonsky [104]

141. M. Drees: "$N = 1$ Supergravity guts with noncanonical kinetic energy terms," Phys. Rev. D **33**, 1468 (1986)

S. Dimopoulos and G.F. Giudice: "Naturalness constraints in supersymmetric theories with nonuniversal soft terms," Phys. Lett. B **357**, 573 (1995)

A. Pomarol and D. Tommasini: "Horizontal symmetries for the supersymmetric flavor problem," Nucl. Phys. B **466**, 3 (1996)

A.G. Cohen, D.B. Kaplan and A.E. Nelson: "The More minimal supersymmetric standard model," Phys. Lett. B **388**, 588 (1996)

142. P. Fayet and J. Iliopoulos: "Spontaneously broken supergauge symmetries and Goldstone spinors," Phys. Lett. **51B**, 461 (1974)

I. Jack, D.R.T. Jones and S. Parsons: "The Fayet-Iliopoulos D-term and its renormalisation in softly-broken supersymmetric theories," Phys. Rev. D **62**, 125022 (2000)

I. Jack and D.R.T. Jones: "The Fayet-Iliopoulos D-term and its renormalisation in the MSSM," Phys. Rev. D **63**, 075010 (2001)

143. G. Dvali and A. Pomarol: "Anomalous $U(1)$ as a mediator of supersymmetry breaking," Phys. Rev. Lett. **77**, 3728 (1996)

A.E. Nelson and D. Wright: "Horizontal, anomalous $U(1)$ symmetry for the more minimal supersymmetric standard model," Phys. Rev. D **56**, 1598 (1997)

T. Barreiro, B. de Carlos, J.A. Casas and J.M. Moreno: "Anomalous $U(1)$, gaugino condensation and supergravity," Phys. Lett. B **445**, 82 (1998)

J. Hisano, K. Kurosawa and Y. Nomura: "Large squark and slepton masses for the first-two generations in the anomalous $U(1)$ SUSY breaking models," Phys. Lett. B **445**, 316 (1999)

Y. Kawamura and T. Kobayashi: "Soft scalar masses in string models with anomalous $U(1)$ symmetry," Phys. Lett. B **375**, 141 (1996); *Erratum-ibid.* B **388**, 867 (1996)

144. D.E. Kaplan, F. Lepeintre, A. Masiero, A.E. Nelson and A. Riotto: "Fermion masses and gauge mediated supersymmetry breaking from a single $U(1)$," Phys. Rev. D **60**, 055003 (1999)

145. H. Cheng, J.L. Feng and N. Polonsky: "Superoblique corrections and nondecoupling of supersymmetry breaking," Phys. Rev. D **56**, 6875 (1997); "Signatures of multi-TeV scale particles in supersymmetric theories," Phys. Rev. D **57**, 152 (1998)

S. Kiyoura, M.M. Nojiri, D.M. Pierce and Y. Yamada: "Radiative corrections to a supersymmetric relation: A new approach," Phys. Rev. D **58**, 075002 (1998)

146. H. Baer, M.A. Diaz, P. Quintana and X. Tata: "Impact of physical principles at very high energy scales on the superparticle mass spectrum," JHEP **0004**, 016 (2000)
H. Baer, C. Balazs, P. Mercadante, X. Tata and Y. Wang: "Viable supersymmetric models with an inverted scalar mass hierarchy at the GUT scale," Phys. Rev. D **63**, 015011 (2001)
H. Baer, C. Balazs, M. Brhlik, P. Mercadante, X. Tata and Y. Wang: "Aspects of supersymmetric models with a radiatively driven inverted mass hierarchy," hep-ph/0102156

147. M. Dine, R. Leigh and A. Kagan: "Flavor symmetries and the problem of squark degeneracy," Phys. Rev. D **48**, 4269 (1993)
Y. Nir and N. Seiberg: "Should squarks be degenerate?," Phys. Lett. B **309**, 337 (1993)
M. Leurer, Y. Nir and N. Seiberg, "Mass matrix models: The Sequel," Nucl. Phys. B **420**, 468 (1994)
S. Dimopoulos, G.F. Giudice and N. Tetradis: "Disoriented and plastic soft terms: A Dynamical solution to the problem of supersymmetric flavor violations," Nucl. Phys. B **454**, 59 (1995)

148. N. Arkani-Hamed, H. Cheng, J.L. Feng and L.J. Hall: "Probing lepton flavor violation at future colliders," Phys. Rev. Lett. **77**, 1937 (1996); "CP violation from slepton oscillations at the LHC and NLC," Nucl. Phys. B **505**, 3 (1997)
N.V. Krasnikov: "Search for flavor lepton number violation in slepton decays at LEP2 and NLC," Phys. Lett. B **388**, 783 (1996)

149. M. Gell-Mann, P. Ramond and R. Slansky: In: *Supergravity, Proceedings of the Workshop, Stony Brook, New York, 1979*, eds. P. van Nieuwenhuizen and D. Freedman (North Holland, Amsterdam 1979) p. 315
T. Yanagida: In: *Proceedings of the Workshop on Unified Theories and Baryon Number in the Universe, Tsukuba, Japan, 1979*, eds. A. Sawada and A. Sugamoto (KEK Report No. 79-18, Tsukuba 1979)

150. K.S. Babu, J.C. Pati and F. Wilczek: "Fermion masses, neutrino oscillations, and proton decay in the light of SuperKamiokande," Nucl. Phys. B **566**, 33 (2000)

151. R.N. Mohapatra: "Theories of neutrino masses and mixings," hep-ph/9910365
S. Raby: 'Neutrino oscillations in a 'predictive' SUSY GUT," hep-ph/9909279

152. F. Vissani and A.Y. Smirnov: "Neutrino masses and $b - \tau$ unification in the supersymmetric standard model," Phys. Lett. B **341**, 173 (1994)
A. Brignole, H. Murayama and R. Rattazzi: "Upper bound on hot dark matter density from $SO(10)$ Yukawa unification," Phys. Lett. B **335**, 345 (1994)
J.A. Casas, J.R. Espinosa, A. Ibarra and I. Navarro: "Neutrinos and gauge unification," Phys. Rev. D **63**, 097302 (2001)

153. H. Nilles and N. Polonsky: "Supersymmetric neutrino masses, R symmetries, and the generalized μ-problem," Nucl. Phys. B **484**, 33 (1997)
N. Polonsky: "The neutrino mass and other possible signals of lepton-number violation in supersymmetric theories," Nucl. Phys. Proc. Suppl. **52A**, 97 (1997)

154. See also comprehensive discussions in:
B.C. Allanach, A. Dedes and H.K. Dreiner: "Two-loop supersymmetric renormalization group equations including R-parity violation and aspects of unification," Phys. Rev. D **60**, 056002 (1999)
M. Bisset, O.C. Kong, C. Macesanu and L.H. Orr: "Supersymmetry without R-parity: Leptonic phenomenology," Phys. Rev. D **62**, 035001 (2000)
S. Davidson and J. Ellis: "Basis-independent measures of R-parity violation," Phys. Lett. B **390**, 210 (1997); "Flavour-dependent and basis-independent

168   References

measures of R violation," Phys. Rev. D **56**, 4182 (1997)
S. Davidson and M. Losada: "Basis independent neutrino masses in the R_P violating MSSM," hep-ph/0010325

155. R. Hempfling: "Neutrino masses and mixing angles in SUSY-GUT theories with explicit R-Parity breaking," Nucl. Phys. B **478**, 3 (1996)

156. M.A. Diaz, J.C. Romao and J.W. Valle: "Minimal supergravity with R-parity breaking," Nucl. Phys. B **524**, 23 (1998)

157. R.M. Godbole, P. Roy and X. Tata: "τ signals of R-parity breaking at LEP200," Nucl. Phys. B **401**, 67 (1993)

158. J. Erler, J.L. Feng and N. Polonsky: "A Wide scalar neutrino resonance and $b\bar{b}$ production at LEP," Phys. Rev. Lett. **78**, 3063 (1997)

159. K. Agashe and M. Graesser: "R-parity violation in flavor changing neutral current processes and top quark decays," Phys. Rev. D **54**, 4445 (1996)

160. L. Navarro, W. Porod and J.W. Valle: "Top-quark phenomenology in models with bilinearly and spontaneously broken R-parity," Phys. Lett. B **459**, 615 (1999)

161. H. Dreiner and G.G. Ross: "R parity violation at hadron colliders," Nucl. Phys. B **365**, 597 (1991)

162. D. Gerdes [CDF and D0 Collaborations]: "Top quark physics results from CDF and D0," hep-ex/9609013

163. S. Dimopoulos, R. Esmailzadeh, L.J. Hall and G.D. Starkman: "Cross-sections for lepton and baryon number violating processes from supersymmetry at $p\bar{p}$ colliders," Phys. Rev. D **41**, 2099 (1990)
H. Dreiner, P. Richardson and M.H. Seymour: "Resonant slepton production in hadron hadron collisions," Phys. Rev. D **63**, 055008 (2001)

164. F. Borzumati, J. Kneur and N. Polonsky: "Higgs-Strahlung and Slepton-Strahlung at hadron colliders," Phys. Rev. D **60**, 115011 (1999)

165. P. Langacker and N. Polonsky: "Implications of Yukawa unification for the Higgs sector in supersymmetric grand unified models," Phys. Rev. D **50**, 2199 (1994)

166. M. Drees, M. Gluck and K. Grassie: "A new class of false vacua in low-energy N=1 supergravity theories," Phys. Lett. B **157**, 164 (1985)

167. M. Claudson, L.J. Hall and I. Hinchliffe: "Low-energy supergravity: False vacua and vacuous predictions," Nucl. Phys. B **228**, 501 (1983)
A. Kusenko, P. Langacker and G. Segre: "Phase transitions and vacuum tunneling into charge and color breaking minima in the MSSM," Phys. Rev. D **54**, 5824 (1996)

168. U. Sarid: "Tools for tunneling," Phys. Rev. D **58**, 085017 (1998)

169. J.F. Gunion, H.E. Haber and M. Sher: "Charge / color breaking minima and A-parameter bounds in supersymmetric models," Nucl. Phys. B **306**, 1 (1988)
J.A. Casas, A. Lleyda and C. Munoz: "Strong constraints on the parameter space of the MSSM from charge and color breaking minima," Nucl. Phys. B **471**, 3 (1996)
H. Baer, M. Brhlik and D. Castano: "Constraints on the minimal supergravity model from non-standard vacua," Phys. Rev. D **54**, 6944 (1996)
A. Kusenko and P. Langacker: "Is the vacuum stable?," Phys. Lett. B **391**, 29 (1997)
S.A. Abel and C.A. Savoy: "On metastability in supersymmetric models," Nucl. Phys. B **532**, 3 (1998)
U. Ellwanger and C. Hugonie: "Constraints from charge and color breaking minima in the (M+1)SSM," Phys. Lett. B **457**, 299 (1999)

170. N. Polonsky and S. Su: "Low-energy limits of theories with two supersymmetries," Phys. Rev. D **63**, 035007 (2001)

171. N. Polonsky: "The scale of supersymmetry breaking as a free parameter," hep-ph/0102196
172. S.P. Martin: "Dimensionless supersymmetry breaking couplings, flat directions, and the origin of intermediate mass scales," Phys. Rev. D **61**, 035004 (2000)
173. K. Hikasa and Y. Nakamura: "Soft breaking correction to hard supersymmetric relations: QCD corrections to squark decay," Z. Phys. C **70**, 139 (1996)
174. H.E. Haber and R. Hempfling, "The renormalization group improved Higgs sector of the minimal supersymmetric model," Phys. Rev. D **48**, 4280 (1993)
175. See the last two references in Ref. [116]
176. F. del Aguila, M. Dugan, B. Grinstein, L. Hall, G.G. Ross and P. West, "Low-energy models with two supersymmetries," Nucl. Phys. B **250**, 225 (1985)

Lecture Notes in Physics

For information about Vols. 1–538
please contact your bookseller or Springer-Verlag

Monographs
For information about Vols. 1–26
please contact your bookseller or Springer-Verlag